THE
SCHOOL
MATHEMATICS
PROJECT

ADDITIONAL
MATHEMATICS BOOK

PART 2

CAMBRIDGE UNIVERSITY PRESS

CAMBRIDGE

LONDON · NEW YORK · MELBOURNE

Published by the Syndics of the Cambridge University Press
The Pitt Building, Trumpington Street, Cambridge CB2 1RP
Bentley House, 200 Euston Road, London NW1 2DB
32 East 57th Street, New York, NY 10022, USA
296 Beaconsfield Parade, Middle Park, Melbourne 3206, Australia

© Cambridge University Press 1968 1970

ISBN 0 521 07877 6

First published 1968
Metricated edition 1970
Reprinted 1972 1974 1977

Printed in Great Britain at the
University Press, Cambridge

THE
SCHOOL MATHEMATICS
PROJECT

When the SMP was founded in 1961, its main objective was to devise radically new secondary-school mathematics courses (and corresponding GCE and CSE syllabuses) to reflect, more adequately than did the traditional syllabuses, the up-to-date nature and usages of mathematics.

This objective has now been realized. SMP *Books 1–5* form a five-year course to the O-level examination 'SMP Mathematics'. *Books 3T, 4* and *5* give a three-year course to the same O-level examination (the earlier *Books T* and *T4* being now regarded as obsolete). *Advanced Mathematics Books 1–4* cover the syllabus for the A-level examination 'SMP Mathematics' with *Revised Advanced Mathematics Books 1* and *2* now available as alternatives to *Advanced Mathematics Books 1* and *2*. Five shorter texts cover the material of the various sections of the A-level examination 'SMP Further Mathematics'. There are two books for 'SMP Additional Mathematics' at O-level. All the SMP GCE examinations are available to schools through any of the Examining Boards.

Books A–H, originally designed for non-GCE streams, cover broadly the same development of mathematics as do the first few books of the O-level series. Most CSE Boards offer appropriate examinations. In practice, this series is being used very widely across all streams of comprehensive schools, and its first seven books, followed by *Books X, Y* and *Z*, provide a course leading to the SMP O-level examination. *SMP Cards I* and *II* provide an alternative treatment of the mathematics in *Books A–D* in card form. *SMP 7–13*, designed for children within that age range, begins publication in 1977.

Teachers' Guides accompany all these series of books.

The SMP has produced many other texts, and teachers are encouraged to obtain each year from the Cambridge University Press, Bentley House, 200 Euston Road, London NW1 2DB the full list of SMP books currently available. In the same way, help and advice may always be sought by teachers from the Executive Director at the SMP Office, Westfield College, Hampstead, London, NW3 7ST, from which may also be obtained the annual Reports, details of forthcoming in-service training courses and so on.

SMP will continue to develop its research into the mathematical curriculum. The team of SMP writers, numbering around a hundred school and university mathematicians, is continually evaluating old work and preparing for new. But at the same time, the effectiveness of the SMP's future work will depend, as it always has done, on obtaining reactions from a wide variety of teachers – and also from pupils – actively concerned in the class-room. Readers of the texts can therefore send their comments to the SMP in the knowledge that they will be warmly welcomed.

October 1976

This part of the Additional Mathematics Book is based on the original contributions of

H. M. Cundy	T. A. Jones
L. E. Ellis	P. G. T. Lewis
D. A. Hobbs	G. D. Stagg
J. S. T. Woolmer	

and has been edited by H. Martyn Cundy assisted by L. E. Ellis.

Many other schoolteachers have been directly involved in the further development and revision of the material and the Project gratefully acknowledges the contributions which they and their schools have made.

CONTENTS

PREFACE

In the preface to the first part of this book the three main uses of Additional Mathematics were set out: for the O-level candidate who has a thorough grasp of the Elementary Mathematics course; for the sixth form Arts student who enjoys mathematics and can afford to spend time exploring some of its developments; and for the sixth form scientist who cannot spare the time for a full A-level course in Mathematics, but who needs to take his study of the subject a stage further for the benefit of his scientific work.

The problem of satisfying these three classes has been felt acutely in this second part. In introducing the calculus it has been considered essential to give an understanding of the notations and techniques in common use, while building on the ideas of function and rate of change in the O-level course and the first part of this book. The treatment adopted here is in a sense alternative to that in the S.M.P. Advanced Mathematics course and bridges over more quickly to the traditional applications of the calculus. Inevitably this means that many results have to be taken on trust with only a sketch of a proof. But the scientist may well prefer the early use of the dy/dx and $\int f(x)\, dx$ notations which is here adopted.

Chapters 11 and 13 introduce the circular and exponential functions in a way which lays emphasis on their physical applications without ignoring the mathematical problems involved. The main routes through some interesting country can be explored in this way without too much distracting detail.

Chapter 14 on vectors recapitulates earlier material and applies it to velocities, forces, and some useful coordinate geometry, concluding with the practical technique of the solution of triangles. This lays the foundation for a study of particle dynamics based on Newton's laws begun in Chapter 15. This is treated vectorially throughout, and emphasis is laid on fundamental principles which are discussed rather more fully than is customary at this stage. Simple ideas of momentum are reached, but not work and energy, which seem to call for an insight best found in the physics laboratory.

The next two chapters develop the statistics content of the course. Once again it has been thought desirable to introduce fresh ideas rapidly; measures of spread and a measure of correlation are defined, the coefficient chosen for the latter being Kendall's coefficient of rank correlation, which is easy to compute and interpret. In the probability chapter the theory is carried sufficiently far to make possible a discussion of the ways in which patterns appear when the results of simple experiments are recorded—the throwing of dice, tossing of coins, and the like. In this way the student is introduced to the binomial and Normal distributions of numerical data.

As in the first part, the final chapter is designed by way of relief from the serious study of the earlier part of the book. In it we take a look at the heavens and our motion in relation to the distant stars. The unifying theme of the chapter is the measurement of time, but we turn aside from this now and again to look at the phases of the moon and the movements of the planets. Trigonometrical techniques acquired earlier are put to good use, and the whole constitutes a fitting reminder to the mathematical student that a universe awaits his contemplation and discovery, vaster by far than a few typographical symbols imprinted on a portion of a Cartesian plane.

ACKNOWLEDGEMENTS

We are greatly indebted to the Cambridge University Press for their patience and co-operation throughout the preparation of this book.

We acknowledge with thanks permission granted by the Oxford and Cambridge Schools Examination Board for the use of questions from the S.M.P. Examination papers.

We wish particularly to record our gratitude to the secretary of the Project, Miss Anne Freeman, and to Mrs Elizabeth Muir and Mrs Henrietta Goggs, who have responded with cheerfulness and efficiency to heavy demands for typing work in the course of production of this book.

A NOTE ON METRICATION

(i) All quantities of money have been expressed in pounds (£) and new pence (p).

(ii) All measures have been expressed in metric units. The fundamental units of the Système International (that is the metric system to be used in Great Britain) are the metre, the kilogram and the second. These units have been used in the book except where practical classroom considerations or an estimation of everyday practice in the years to come have suggested otherwise.

(iii) The notation used for the abbreviations of units and on some other occasions conforms to that suggested in the British Standard publications PD 5686: 1967 and BS 1991: Part 1: 1967.

9

DIFFERENTIATION

1. RATE OF CHANGE

1.1 Acceleration and speed. Consider the statement made by a defending motorist—'I accelerated uniformly from 40 km/h to 60 km/h passing the de-restriction sign at exactly 50 km/h'.

What does he mean by 'accelerated uniformly'? Or, even more basically, what does he mean at a speed of 50 km/h?

The idea of a constant speed is relatively easy to grasp; if a car is going at a constant speed of 50 km/h then we know that it will steadily cover 1 km every 1·2 minutes. But what of variable speed? What does the motorist mean by saying that he passed the de-restriction sign at 50 km/h? Suppose that as he passed the sign we had a very accurate instrument that could measure the distance covered in very small intervals of time; what would we expect to find? One would hope to find that, as we considered smaller and smaller intervals, his average speed would become steadily closer to 50 km/h. We could not of course consider an interval of zero time: without a definite interval of time there is no definite distance travelled and hence no average speed. Thus if we are to formulate a meaningful definition it will have to be based on distances covered in small intervals of time. We shall consider this idea more fully below.

In the case of uniform acceleration there is much less difficulty. The acceleration, by definition, is the same throughout the interval; and thus if the car took 10 seconds to accelerate from 40 km/h to 60 km/h we could say with confidence that its acceleration was 2 km/h per second (or $\frac{5}{9}$ m/s²) throughout the interval. If the acceleration had not been uniform we should have experienced difficulties similar to those in the paragraph above.

1.2 Rates of change and gradient. Figure 1 is a plan of the Rheims motor racing circuit where the French Grand Prix is often held. Let us consider the sort of progress that a driver might make. On the first lap he would accelerate past the pits, brake slightly at the first right-hand bend, build up to nearly top speed until a check for the curve before Muizon, and then a further check at the hairpin itself. Follow the circuit round through the Thillois corner and the second (flying start) lap for yourself.

Figure 2 gives a possible graph of speed against time. When were the hairpin bends taken? On what part of the circuit did the driver accelerate fastest? Why does the graph go down more sharply than it comes up?

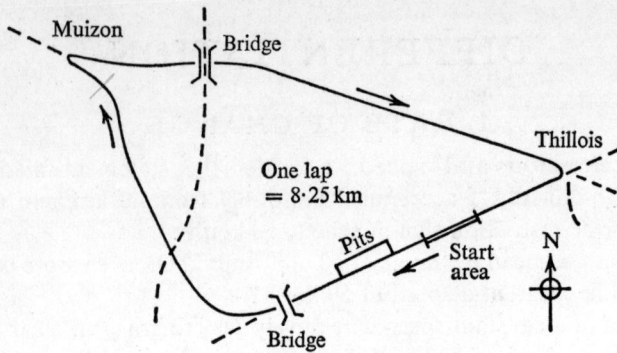

Fig. 1. Rheims circuit.

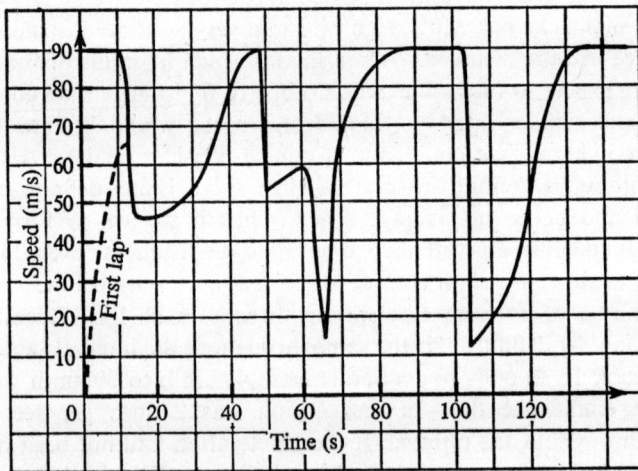

Fig. 2. Speed-time graph–Rheims.

Let us examine the part of the graph representing the situation about a minute after the lap has started (Figure 3). In particular we will consider the 10-second interval during which the driver breaks sharply into the Muizon corner and accelerates out of it. One could say that over the whole period his speed increases by 12 m/s and thus that his *average acceleration* is 1·2 m/s². This would give a very misleading impression of his progress and this illustrates a very important point: when dealing with the way in which quantities are changing we must take a small interval and look at the 'local' situation, otherwise we shall get some very distorted results.

194

Actually for the first 3 seconds the driver decelerates rapidly from 58 m/s to 16 m/s, thus losing speed at a rate of 14 m/s². Similarly, over the next 7 seconds he accelerates at a roughly constant rate of about 8 m/s².

In Figure 3, what do the gradients of the lines *AB, BC, AC* represent?

How justified are we in saying that in the 3-second interval he decelerated at 14 m/s²? Would we be more justified (and perhaps able to give the result to a second significant figure) if we knew his speed accurately at a number of instants in between? Of course, whatever time interval we take

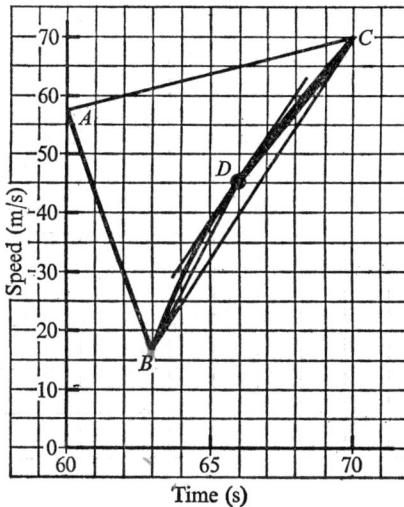

Fig. 3

over which to calculate the rate at which the speed is changing, what we shall really find is the average acceleration over that interval however small or large it may be. On the speed-time graph the gradient of the line joining two points is the average acceleration over the interval which separates them, the unit of acceleration being the unit of speed divided by the unit of time. Considerations like these lead us to define the acceleration as the gradient of the speed-time graph.†

For instance, suppose we wanted to find the actual acceleration at the instant represented by the point *D* in Figure 3. It would obviously be grossly misleading to estimate this acceleration from the gradient of the chord *AC*; a better estimate could be obtained by considering the gradient of *BD, DC,* or *BC*. To improve on this still further, we could take a point

† Strictly, as we shall see in Chapter 15, acceleration is a vector quantity; what we are talking about here is its component along the track. There is, unfortunately, no word for 'rate of change of speed'.

B_1 on the speed-time curve between B and D and find the gradient of the chord $B_1 D$; then a point B_2 between B_1 and D and find the gradient of the chord $B_2 D$, and so on. We see intuitively that in the neighbourhood of a 'smooth' point such as D the gradients of these chords steadily approach the gradient of the tangent at D, which represents the acceleration at this point.

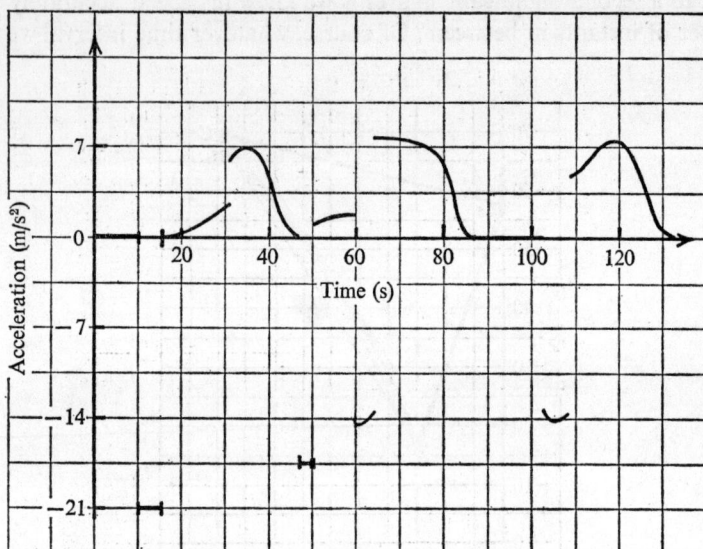

Fig. 4

On the other hand, the definition of the acceleration at a point such as B presents problems; indeed, in the strict sense it cannot be defined. If we consider time-intervals ending at B, we find a deceleration at B given by the gradient of the left-hand branch of the curve (i.e. as the car approaches the corner): but equally well if we consider time-intervals beginning at B we obtain a totally different answer.

In fact unless the answers obtained by applying a limiting process to the gradients of chords to the left and to the right of the point we are considering are the same, we say that the gradient of the curve is undefined.

This is shown in Figure 4 by the fact that the acceleration-time graph is discontinuous. Figure 4 gives a rough sketch of the acceleration-time graph for Figure 2. To summarize: acceleration measures the rate at which speed is changing and may be found by considering the gradient of the tangent to the curve at the appropriate point of the speed-time graph. Speed, itself, is the rate at which distance (in this case the distance travelled along the circuit) is changing.

Exercise A

1. At what other points is the acceleration undefined?

2. When is the acceleration zero? At what speed does this occur?

3. Can you think of any physical situation in which the speed of a particle might be undefined? Is the speed of the car in the last section always defined?

4. Sketch the graph of the gradient of the following graph (Figure 5).

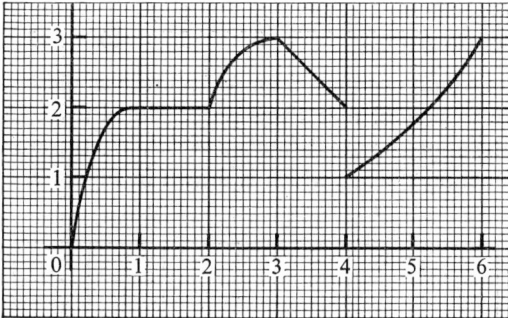

Fig. 5

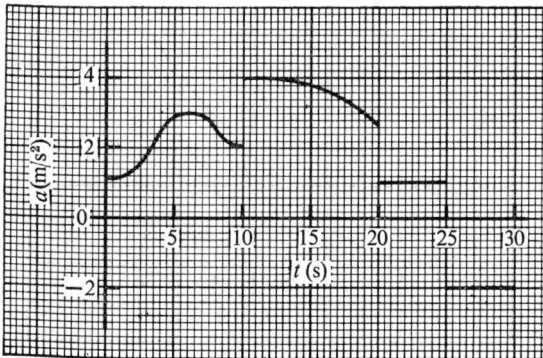

Fig. 6

5. The following readings were taken of a car's speedometer:

Time after starting (seconds)	0	2	4	6	8	10
Speed (m/s)	0	2	4	7	9	12

Plot a speed-time graph, and estimate
(a) the average acceleration over the first 5 seconds;
(b) the acceleration at the end of 5 seconds.

6. Figure 6 shows the acceleration-time graph for a go-kart starting from rest. Sketch the shape of the speed-time graph.

197

7. Draw a plan of a bicycle grand prix circuit using roads or paths in your school grounds. Sketch a speed-time graph for it. (Make it simple! If you want a hard example, find a plan of Nürburgring.)

2. GRADIENT FUNCTION

2.1 Speed of a falling stone. Galileo once observed a stone falling from the top of a convenient building in Pisa. We could repeat this experiment today and obtain results like these:

Time after dropping (in seconds)	0	1	2	3	4	5
Distance fallen (in metres)	0	5	20	45	80	125

Galileo's results would not have been so accurate as these, but were sufficiently accurate to suggest that s was proportional to t^2. In our units we would obtain the approximate formula $s = 5t^2$. We therefore examine the graph of the function $t \rightarrow 5t^2$, which is drawn in Figure 7.

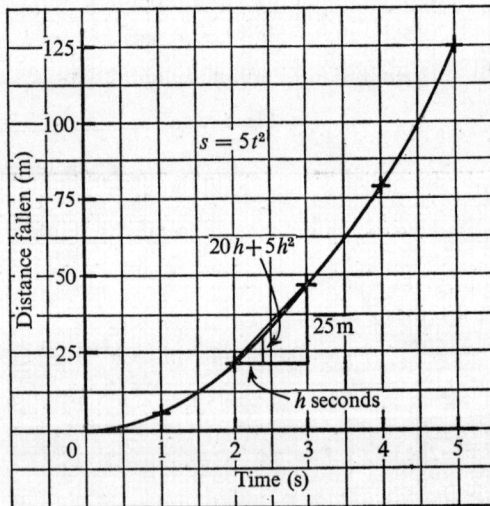

Fig. 7

How fast is the stone falling after 2 seconds? From the table we find that in the first 2 seconds it fell 20 m, and in the next second it fell a further 25 m. Thus the average speed for the first 2 seconds is 10 m/s and for the third second is 25 m/s.

To get a closer estimate, we will look at the interval of time from when $t = 2$ to $t = 2+h$. In this interval the distance fallen by the stone is

$$5(2+h)^2 - 5.2^2 \text{ m}$$

$$= 20h + 5h^2 \text{ m}.$$

Thus the average speed in m/s in this interval of time h seconds is given by $v = 20+5h$. If we put $h = 1$ in this formula we obtain $r = 25$ m/s as the average speed from $t = 2$ to $t = 3$, in accordance with the result we have obtained from the table. But we can now obtain additional information. Putting $h = 0.1$ we obtain 20.5 m/s for the average speed from $t = 2$ to $t = 2.1$; putting $h = 0.01$ we obtain 20.05 m/s for the average speed from $t = 2$ to $t = 2.01$, and so on. We may tabulate this as follows:

Time interval h seconds	Average speed in m/s
1	25
0·1	20·5
0·01	20·05
0·001	20·005
...	...

It is clear that the average speed is tending to 20 m/s as the time interval shrinks to zero, and indeed this can be shown rigorously from the formula

$$v = 20+5h.$$

Exercise. Repeat the work above to find the speed v when $t = \frac{1}{2}$, 1, and 3, and comment on your results.

2.2   Derived function. For each instant of time there must be a definite speed v m/s, and this is given by the gradient of the graph of the function $t \to 5t^2$. We have now found this gradient for four values of t, and it can in principle be found for every value. The value of the gradient at any point we shall call the *derivative* of f at that point. Thus we have found the derivative of $f: t \to 5t^2$ at $t = 2$ to be 20. The function which maps t onto the derivative of f at t is called the *derived function* of f and is written f'. Thus $f'(2) = 20$.

To find the derived function we adopt a general method, as follows. The average speed in the interval from t to $t+h$ is given by

$$\frac{5(t+h)^2 - 5t^2}{h} = \frac{5(t^2+2th+h^2-t^2)}{h}$$

$$= 10t+5h$$

and this tends to $10t$ as h tends to zero. We conclude that

$$f(t) = 5t^2 \Rightarrow f'(t) = 10t.$$

Definition 1. $f'(t)$ is the limit as h tends to zero of

$$\frac{f(t+h)-f(t)}{h}.$$

There are two points to be noted about this. First, we cannot simply put $h = 0$ in the formula for the average speed, since average speed over a zero

199

time interval is meaningless. We can, however, say that the average speed can be made as near to $10t$ as we like by taking h sufficiently small, and this is what we mean when we say that the average speed tends to $10t$ and that this is the speed at the instant t.

Secondly, it is worth noting that if we take a time interval from $t-h$ to $t+h$, *surrounding* the point t, we obtain for the average speed over this interval

$$\frac{5(t+h)^2-5(t-h)^2}{2h},$$

and this is equal to $10t$ whatever the value of h. This happens in this case because of certain geometrical properties of the parabola, but for many other functions the use of a symmetrical interval leads to simpler algebra. Formally we can then state

Definition 2. $f'(t)$ is the limit as h tends to zero of

$$\frac{f(t+h)-f(t-h)}{2h}.$$

These two definitions can be proved to be equivalent, provided the limit of Definition 1 exists; we shall use whichever is convenient.

2.3   Gradient functions of the functions $t \to t^3$ and $t \to t^{-1}$. For the function $t \to t^3$ (Figure 8) the average gradient shown is

$$\frac{(t+h)^3-(t-h)^3}{2h}$$

$$=\frac{(t^3+3t^2h+3th^2+h^3)-(t^3-3t^2h+3th^2-h^3)}{2h}$$

$$=\frac{(6t^2h+2h^3)}{2h}$$

$$=3t^2+h^2.$$

From this result it is obvious that as we take smaller values for h, the gradient of the chord tends to $3t^2$.

Thus $f(t) = t^3 \Rightarrow f'(t) = 3t^2.$

For the function $t \to t^{-1}$ (Figure 9) the gradient of the chord is

$$\frac{\dfrac{1}{t+h}-\dfrac{1}{t-h}}{2h}$$

$$=\frac{-2h}{2h(t^2-h^2)}$$

$$=\frac{-1}{t^2-h^2}.$$

Clearly as h tends to 0 this tends to $-1/t^2$, or $-t^{-2}$.

Hence $f(t) = t^{-1} \Rightarrow f'(t) = -t^{-2} \quad (t \neq 0).$

200

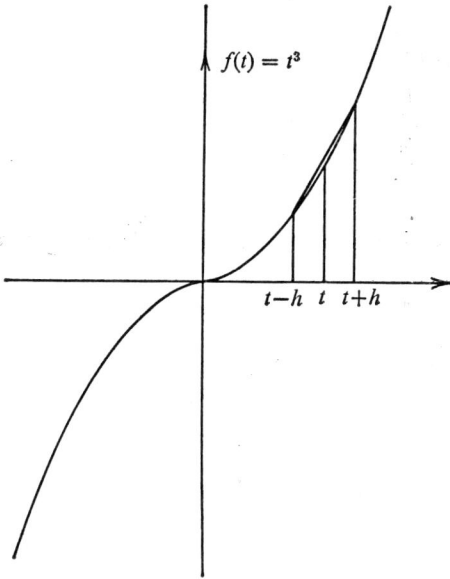

Fig. 8

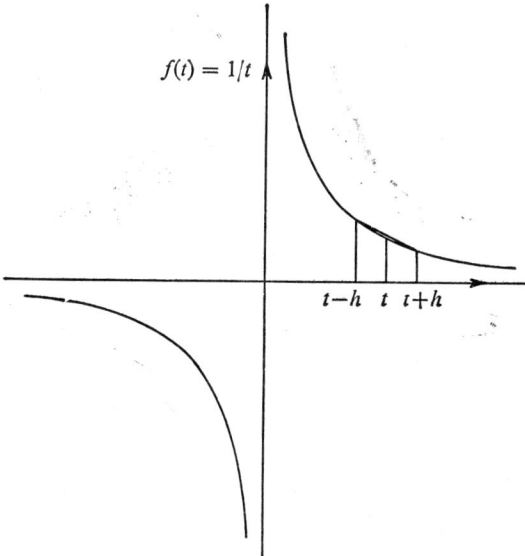

Fig. 9

Exercise B

1. Find $f'(t)$, using definition 1, if $f(t)$ is

    (a) t;   (b) $3t$;   (c) $3t+2$;   (d) $at+b$.

2. From definition 1, find $f'(t)$ if $f(t)$ is

    (a) t^2;   (b) $3t^2+4$;   (c) $3t^2+5t$;   (d) $3t^2+5t+4$.

3. What is the acceleration of the falling stone described at the beginning of Section 2?

4. Use definition 2 to find $f'(t)$ if $f(t) = t^3 + 3t^2 + 7$.

5. A particle is moving in a straight line so that its distance, s metres, from a fixed point to the line t seconds after starting, is given by

$$s = \tfrac{1}{2}t^2 - t.$$

Find its speed and acceleration at any time.

6. If $f: t \to t^{-1}$, explain clearly why you would expect $f'(t)$ to be everywhere negative.

*7. Prove that
$$\frac{a^4 - b^4}{a - b} = (a+b)(a^2 + b^2).$$

Deduce that
$$\frac{(t+h)^4 - (t-h)^4}{2h} = 4t(t^2 + h^2).$$

What do you conclude to be the derived function of $f: t \to t^4$?

*8. Find the derived function of $f: t \to 1/t^2$.

3. DIFFERENTIATION

The process of finding the derived function is often called *differentiation*; the word is not very happily chosen, and 'derivation' would be preferable, but it is well established in use and we shall conform to this. Although we have so far considered only rates of change with respect to time, and have therefore used t as the name of a typical element of the domain of a function f, there is no reason why any other letter would not serve equally well. Thus,   if $f: t \to t^2$,   then $f'(t) = 2t$,

but also $f: x \to x^2$, $f'(x) = 2x, f'(u) = 2u$, and so on.

3.1 Derivatives of powers of x. If we tabulate the results we have obtained in Exercise B, a pattern begins to appear, as the following table shows:

202

$f(x)$	$f'(x)$
x^4	$4x^3$
x^3	$3x^2$
x^2	$2x$
x	1
1	0
x^{-1}	$-x^{-2}$
x^{-2}	$-2x^{-3}$

These are all particular cases of a general rule, which makes them easy to remember, namely
$$f(x) = x^n \Rightarrow f'(x) = nx^{n-1}.$$

We shall not prove this here, though in fact it is true for all rational numbers n.

Exercise C

1. Draw accurately on the same axes the graphs of $f(x)$, $g(x)$ and $s(x)$, where $-3 \leqslant x \leqslant 3$, and
$$f: x \to x,$$
$$g: x \to x^2,$$
$$s: x \to x+x^2.$$

By drawing and measurement, complete the following table. What do you conclude?

x	-2	-1	0	1	2
$f'(x)$					
$g'(x)$					
$s'(x)$					

2. Draw accurately on the same axes the graphs of $f(x)$ and $g(x)$ where $-3 \leqslant x \leqslant 3$, and
$$f: x \to x^3,$$
$$g: x \to \tfrac{2}{3}x^3.$$

By drawing tangents and measuring their gradients, complete the following table. What do you conclude?

x	-2	-1	0	1	2
$f'(x)$					
$g'(x)$					

These two exercises suggest two results which are easily proved, if we assume the corresponding results for limits.

A. If $f(x)+g(x) = s(x)$, then $f'(x)+g'(x) = s'(x)$. We have

$$\frac{f(x+h)-f(x)}{h}+\frac{g(x+h)-g(x)}{h} = \frac{s(x+h)-s(x)}{h},$$

and the result follows on letting h tend to zero.

B. If $g(x) = kf(x)$ where k is a number, then $g'(x) = kf'(x)$. This again follows immediately from the fact that

$$\frac{g(x+h)-g(x)}{h} = \frac{kf(x+h)-kf(x)}{h} = k\cdot\frac{f(x+h)-f(x)}{h}.$$

3. Use the rules given above to write down the derived functions of the following:

(a) $f: x \to 7x^6-3x^3$; (b) $f: x \to 5x^2-4x+3$;

(c) $f: t \to 16t^2+t-5$; (d) $f: u \to (12/u)$;

(e) $f: p \to 1-(1/p^2)$; (f) $f: t \to t^4+4t^2+4$.

4. Find the derived functions, and the gradients when $x = 2$, of the following functions:

(a) $x \to 1-x$; (b) $x \to 1-x^2$;

(c) $x \to x^2+3x-2$; (d) $x \to x^2+(1/x^2)$.

5. A cricket ball is thrown up so that its height, h metres, above the ground t seconds after starting, is given by

$$h = 5+12\cdot5t-5t^2.$$

Write down the speed of the ball: (a) after t seconds; (b) after 1 second; (c) after $1\frac{1}{4}$ seconds.

Explain your answer to (c).

6. What is the acceleration of the ball in Question 5? (Use your answer to 5(a). Can you explain the sign of the acceleration?)

3.2 A new notation. Let us return for a moment to the falling stone of Section 2.1. If it falls s metres in t seconds, then approximately

$$s = 5t^2.$$

This is a *formula* giving s in terms of t, and expresses a function mapping t onto s. Hitherto we have used a letter to denote this function and have written $s = f(t)$ where $f: t \to 5t^2$. This is the mathematician's point of view: he is interested in the functional relationship between t and s, and once this is known he can say that the derived function is

$$f': t \to 10t.$$

This, of course, gives the speed at time t, which is all-important to the physicist, but there is nothing in the notation to indicate this.

For this purpose we need a new notation which will direct attention, not to the functional relationship, but to the quantities which are being

204

related—in this case to s and t. We wish to indicate that $10t$ is the rate of change of s with respect to t, and to do this we write

$$\frac{ds}{dt} = 10t.$$

We read the left-hand side of this equation as 'ds by dt', and it means 'the rate of change of s with respect to t'. To evaluate it we should need to have recourse to the function mapping t onto s, but we need not give this function a name. Instead we may proceed direct from the formula

$$s = 5t^2$$

to the rate of change $$\frac{ds}{dt} = 10t.$$

We can use this notation whatever the letter-names of the quantities related by the function. Thus:

$$s = f(t) \Rightarrow \frac{ds}{dt} = f'(t),$$

$$y = f(x) \Rightarrow \frac{dy}{dx} = f'(x),$$

$$p = f(v) \Rightarrow \frac{dp}{dv} = f'(v)$$

and so on.

[This notation has arisen because, in the historical development of the calculus, as this part of mathematics is called, δs is used to denote the small change is s corresponding to a small change δt in t. Thus $f'(t)$ is the limit of $\delta s/\delta t$ as δt tends to zero, and this is written ds/dt.]

The usefulness of such a notation is seen as soon as we consider the possibility of expressing the speed of the stone in terms of the distance it has fallen, rather than the time. We shall find in fact that $v = \sqrt{20s}$. Now this is a totally different formula from $v = 10t$, and the functions $t \to 10t$ and $s \to \sqrt{20s}$ are quite distinct. Nevertheless, the equations

$$\frac{ds}{dt} = 10t$$

and $$\frac{ds}{dt} = \sqrt{20s}$$

are both true, since the left-hand sides represent the same physical quantity, namely the speed of the falling stone.

3.3 Physical applications. A metal ball is being heated. At any moment during this process it will have a definite volume, V cm³, a definite radius r cm, and a definite temperature, $\theta°$ C. Given the temperature, the radius and volume are defined. Hence there are functions f and g such that

$$r = f(\theta) \quad \text{and} \quad V = g(\theta).$$

205

We may assume that these functions are such that derived functions exist. Accordingly

$$\frac{dr}{d\theta} = f'(\theta)$$

and

$$\frac{dV}{d\theta} = g'(\theta).$$

What are the meanings of the left-hand sides of these equations? Evidently $dr/d\theta$ means 'the rate of change of radius with temperature', and $dV/d\theta$ means 'the rate of change of volume with temperature'.

Suppose, for example, that f is the function

$$f: \theta \to 3 + 0\cdot00002\theta.$$

Then we have

$$r = 3 + 0\cdot00002\theta$$

and

$$\frac{dr}{d\theta} = 0\cdot0002.$$

This means that the radius increases uniformly with temperature at the rate of $0\cdot0002$ cm/degC. In addition, θ might be given in terms of the time, t seconds, by

$$\theta = 18 + 0\cdot4t,$$

so that

$$\frac{d\theta}{dt} = 0\cdot4.$$

This means that the rate of increase of temperature with time is $0\cdot4$ degC/s. dr/dt would then also have a meaning: the rate at which the radius is growing. To find it, we have

$$r = 3 + 0\cdot0002(18 + 0\cdot4t)$$

$$= 3\cdot0036 + 0\cdot00008t,$$

so that $dr/dt = 0\cdot00008$. The radius is growing at the steady rate of 8×10^{-5} cm/s.

Exercise D

Write the statements of Questions 1–7 in mathematical notation.

1. The rate of increase of the pressure p of gas in a cylinder is inversely proportional to the square of its volume v.

2. The faster a car goes, the less it can accelerate; (that is, its acceleration is inversely proportional to its speed).

3. The rate of decrease of atmospheric pressure (n N/cm²) with height (h metres) above sea-level is proportional to the pressure.

4. The rate of decrease of temperature (θ °C) of a cooling body is proportional to the excess of its temperature above that of its surroundings (10 °C).

5. In a certain factory, the rate at which the cost ($£C$) of producing n ball-bearings per day increases with n is inversely proportional to n.

6. The rate of increase of area of a circular ink-blot with respect to the radius is equal to the circumference.

7. The rate at which the tension in a spring increases with the length of the spring is constant.

8. If $V = \frac{4}{3}\pi r^3$, what is dV/dr? Interpret this.

9. If there were N people in Bristol t years after 1900, then $dN/dt = 30(100-t)$. Interpret this in words, and point out any deficiencies in the statement.

10. A particle moves in a straight line so that $s = t^3 - 2t^2$. Investigate its motion for $0 \leqslant t \leqslant 3$, and find when its speed is greatest.

11. Repeat Question 10 for $s = 2t^4 - t$. [$0 \leqslant t \leqslant 2$.]

12. The volume V cm³ of water in a vessel after t seconds is given by

$$V = 1 - 3t + 3t^2.$$

What is the rate of change of volume with respect to time after 3 seconds? What is it after $\frac{1}{2}$ second? What is the significance of this result? Sketch the graph of V against t.

13. The radius r cm of a blot of ink is 2 cm at $t = 0$ and is growing steadily at $\frac{1}{2}$ cm/s. Write down a formula for r after t seconds. Express the area A of the blot in terms of t and find how fast the area is growing after 3 seconds.

14. According to Stefan's law of radiation, the energy E radiated by a 'black body' at temperature T is given by
$$E = kT^4,$$

where k is a constant. What is the rate of change of energy with respect to temperature?

15. The revenue £R from the sale of x articles produced in a factory is given by $R = 100x - 2x^2$. What is the change of revenue with respect to the number of articles produced? Sketch the graph of R against x. What is the maximum revenue that the factory can obtain? Did you use your result for dR/dx to obtain this?

16. The solubility, that is the number of grams dissolved by 100 gm of water, of potassium chlorate at $T°$ C is given by

$$S = 3 + 0\cdot1T + 0\cdot0044T^2.$$

Calculate the rate of change of solubility with temperature at 80 °C.

17. Write down the volume of a cube of side x cm. What is the rate of change of volume with respect to the length of a side? What is the rate of change of the surface area with respect to the length of a side?

18. If $y = x + 1/x$, sketch the graph of y and find dy/dx.

19. If $y = (4x+5)^2$, sketch the graph of y, and by multiplying out the bracket find dy/dx. Is your result similar to the standard result for differentiating x^2?

20. A car takes 20 seconds to accelerate from rest to its maximum speed and its speed is given by
$$v = \tfrac{1}{2}(40t - t^2) \quad (0 \leqslant t \leqslant 20).$$

Sketch the graph of v against t, and find the car's acceleration at any time t. The readings for the car's distance away from the start satisfy an equation of the form $s = at^2 + bt^3$. Find a and b if this distance function gives the correct speed.

21. The table of braking distances for motor vehicles as given in the *Highway Code* is similar to the following:

Speed v km/h	Overall stopping distance s metres
20	6
40	16
60	30
80	48
100	70

Plot this information on a graph. Draw a smooth curve through the points and use it to estimate the rate of change of braking distance with speed of 60 km/h.

*22. The data in Question 21 is in fact consistent with an equation of the form $s = av + bv^2$. Find a and b and hence find ds/dv when $v = 30$. Check your answer to Question 21.

4. SOME MATHEMATICAL APPLICATIONS

4.1 Turning values

Example 1. An open rectangular tank is to have a square base and is to be constructed from a sheet of iron of total area 1200 square metres. What is its greatest possible volume? (Figure 10.)

Suppose that the base is of length x metres and the height is h metres. Then $x^2 + 4xh = 1200$, and $V = x^2 h$.

Using the relation between x and h, we can obtain a formula for V in terms of x and this turns out to define a simple function.

$$h = \frac{1200 - x^2}{4x} \quad \text{and} \quad V = \tfrac{1}{4}x(1200 - x^2).$$

We then make a rough sketch of the graph of this function (Figure 11). This should always be done as it gives a clear indication of exactly what the problem involves.

From the sketch it is clear that for positive x the maximum value of V occurs when the tangent is horizontal. This occurs when $dV/dx = 0$.

Now $dV/dx = 300 - \tfrac{3}{4}x^2$, and so $dV/dx = 0$ if $x = 20$ (the other value -20, is irrelevant). Thus the greatest volume is 4000 m³.

The solution of a wide variety of problems involves finding the greatest and least values taken by functions over given domains.

In Figure 12, over the domain $0 \leqslant x \leqslant 4$, f takes its greatest value at C and its least value at B. Over the domain $0 \leqslant x \leqslant 3$ the greatest value is at A and the least value at B. Over the domain $0 \leqslant x \leqslant 2$ the greatest value is still at A, but the least value is zero. The point A is called a *maximum* point, and the value of f there is a *maximum* value. It should be carefully noted that it is only a 'local' maximum; the greatest value over a given domain may be quite different.

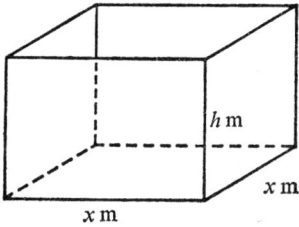

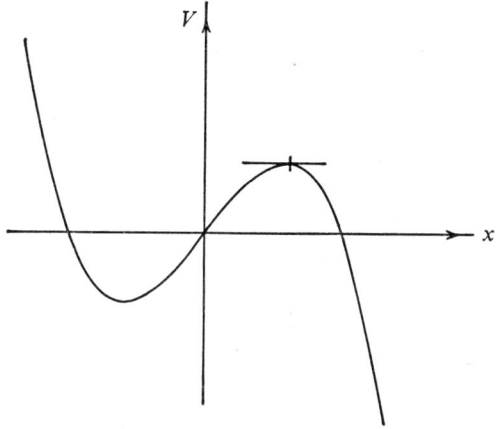

Fig. 10   Fig. 11

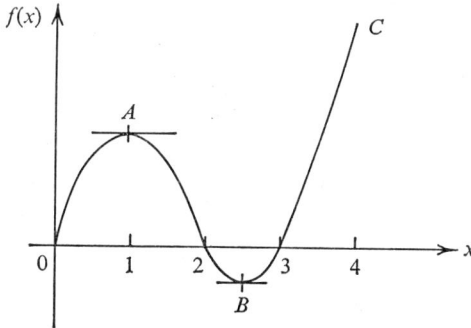

Fig. 12

In the same way B is a *minimum* point, and the value of f at B is a *minimum* value. Points such as A and B are called *turning points* and the corresponding values of f are called *turning values*. At turning values the derivative $f'(x)$ (if it exists) is zero, but not conversely, as Figure 13 shows. At such a point as D, $f'(x) = 0$, but D is not a turning point. It is called a *stationary point of inflexion.*

209

Again a function such as $|x|$ (modulus of x) whose graph is shown in Figure 14 has no derivative at its turning point. The best way to sort out the cases is to draw a graph and if necessary to examine the sign of $f'(x)$.

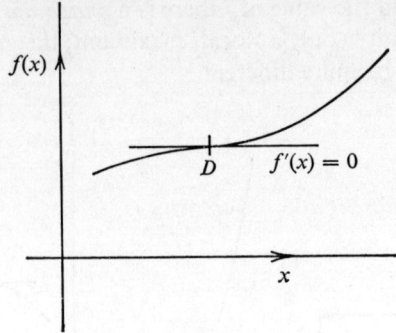

Fig. 13

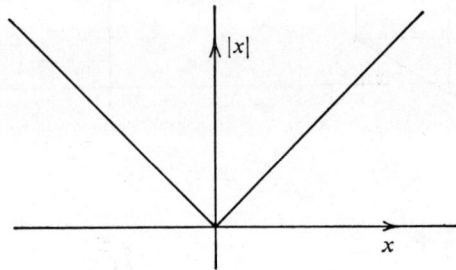

Fig. 14

Example 2. Find the greatest and least values of the function

$$f: x \to 2+x^3-\tfrac{3}{4}x^4 \quad \text{over the domain} \ -1 \leqslant x \leqslant 2.$$

$$f(x) = 2+x^3-\tfrac{3}{4}x^4$$

$$\Rightarrow f'(x) = 3x^2-3x^3$$

$$= 3x^2(1-x).$$

Hence we have the table:

x	-1		0		1		2
$f(x)$	$\tfrac{1}{4}$		2		$2\tfrac{1}{4}$		-2
$f'(x)$	6	$+$	0	$+$	0	$-$	-12

where the blank spaces indicate intermediate values.

210

The graph is shown in Figure 15, from which it is clear that:

(a) the greatest value is $2\frac{1}{4}$ at $x = 1$;

(b) the least value is -2 at $x = 2$;

(c) $x = 0$ gives an inflexion and not a turning value.

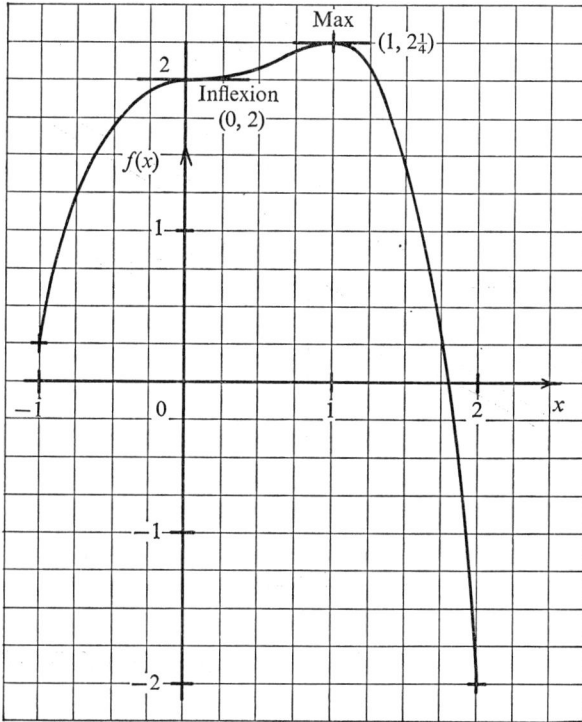

Fig. 15

Example 3. A nut of mass $\frac{1}{4}$ kg is screwed at a distance of x m from the centre of a rod of length 3 m and mass 4 kg. The whole system is set in motion, oscillating about the centre of the rod in a vertical plane. It can be shown that the square of the period of small oscillations is

$$\pi^2(3+\tfrac{1}{4}x^2)/2x.$$

Find the minimum period.

It will clearly be sufficient to consider the function $x \rightarrow (3+\tfrac{1}{4}x^2)/2x$. We refer to the sketch of its graph and find any turning values (see Figure 16).

$$f(x) = 3/2x + x/8.$$

The graph is most quickly sketched by first sketching the graphs of

$$x \rightarrow \frac{x}{8} \quad \text{and} \quad x \rightarrow \frac{3}{2x}$$

and adding their heights. Then $f'(x) = -3/2x^2 + 1/8$, and so there are turning values when $x = \pm\sqrt{12}$. But care is needed as the relevant domain of x is only $0 \leqslant x \leqslant 1\frac{1}{2}$. The least value of the function in this domain will occur when $x = 1\frac{1}{2}$. The corresponding square of the period is $19\pi^2/16$.

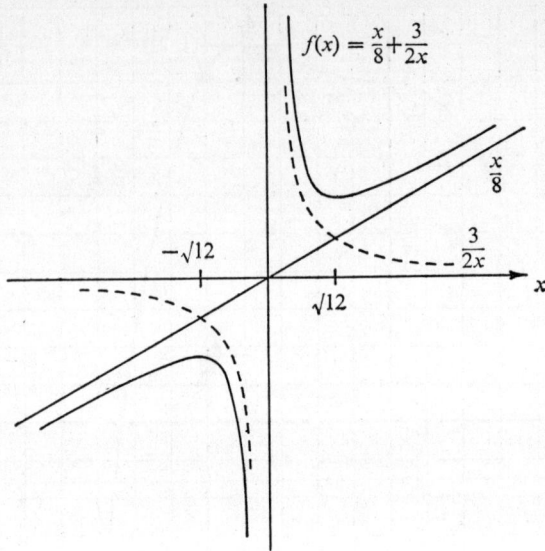

$$f(x) = \frac{x}{8} + \frac{3}{2x}$$

Fig. 16

4.2 Tangents to a curve.

A tangent to a curve at a given point is a straight line satisfying two conditions:
  (i) it passes through the point;
  (ii) it has the same gradient as the curve at the point.

Example 4. To find the equation of the tangent to $y = x^2 - x$ at the point $(3, 6)$. Here $dy/dx = 2x - 1$. Hence the curve has gradient 5 when $x = 3$. Now the equation of a straight line passing through $(3, 6)$ with gradient m is

$$y - 6 = m(x - 3).$$

Hence the required tangent has equation

$$y - 6 = 5(x - 3)$$

or

$$y - 5x + 9 = 0.$$

More generally we find the equation of the tangent to $y = f(x)$ at the point $(a, f(a))$. The curve has gradient $f'(a)$ at $x = a$; and the equation is

$$y - f(a) = f'(a)(x - a)$$

or

$$y = f'(a)x + f(a) - af'(a).$$

212

This is the *linear approximation* to the curve at the point $x = a$, because it is the best-fitting straight line at the point. Note that this is not to say that it is the best fitting straight line in any interval containing the point. The idea can be extended to quadratic approximations (which are the best fitting quadratic curves) and so on, giving a practical method of getting steadily closer approximations to many functions.

Example 5. Find an approximate value for $(1 \cdot 02)^3$.

Let $f(x) = x^3$.

Then $f'(x) = 3x^2$ and $f'(1) = 3$; the linear approximation at $x = 1$ is $f(x) \simeq 1 + 3(x-1)$.

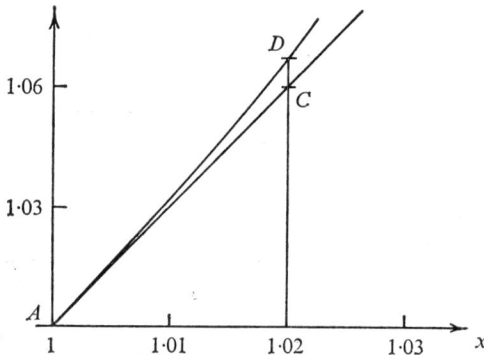

Fig. 17. Sketch of the neighbourhood of $x = 1$.

Hence $f(1 \cdot 02) \simeq 1 \cdot 06$. The actual value is $1 \cdot 061208$, and the error is represented by CD on the sketch (Figure 17).

Miscellaneous Exercise

1. The acceleration a m/s² of a particle moving on a straight line is given after t seconds by

$$a = 3t(3-t).$$

Sketch the graph of a against t and find when the acceleration is greatest. Comment on any symmetry that the graph has.

2. The perimeter of a rectangle is 12 cm. What is its greatest possible area?

3. An open rectangular tank with a square base is to be constructed with fixed surface area A m² and its volume is to be as large as possible. Prove that the height must be half the side of the square base.

4. In order that a parcel may be sent through the post it must not be longer than one metre nor have the sum of its girth (that is, the distance round) and its length more than 2 metres. What is the greatest permissible volume of the parcel (*a*) if it has a square cross-section; (*b*) if it is cylindrical?

213

5. A box is to be made from a sheet of cardboard 1 metre square by cutting away squares from each corner, and then folding along the dotted lines (see Figure 18). Find the size of the squares for the volume to be greatest.

Fig. 18

6. Sketch the graph and find the greatest value of the function
$$f: x \to (x-1)(2-x).$$

7. $y = x^3 + 3x^2 - 9x + 5$; for what values of x between -4 and $+2$ do the greatest and least values of y occur?

8. (a) Find the equation of the tangent to $y = x + x^3$ at the point (2, 10).

(b) Repeat for
$$y = 1 + x + x^2 + x^3 \quad \text{at} \quad x = 3.$$
and for
$$y = x^{-1} \quad \text{at} \quad x = 4 \quad \text{and} \quad x = -4.$$

9. Use the idea of linear approximation to estimate $(5 \cdot 03)^4$.

10. Draw the graph of $x^2 - 5$ and by drawing the tangent at $x = 2$ obtain an estimate of $\sqrt{5}$.

11. I performed a petrol check on my car. The results were as follows:

Litres used	1	2	3	4	5	6	7	8	9	10
km gone	12	24	36	50	62	75	87	101	116	132
Litres used	11	12	13	14	15	16	17	18	19	20
km gone	148	164	180	197	211	228	246	263	279	295

Draw a graph of distance travelled against petrol used. Roughly, what were my greatest and least rates of consumption? Is there any significant change during the test?

12. Differentiate $x(x-1)(x-2)$ and sketch the graph of the function. Can you see where its point of inflexion is? (See note at the end of the exercise.)

13. The area of a rectangular cattle pen is to be made as large as possible. 60 metres of fencing are available. What is the greatest possible area that can be enclosed? Answer the same problem if on one side of the rectangle a hedge can be used as a boundary.

14. Evaluate approximately the polynomial x^3+5x^2-7x+3 when $x = 1\cdot05$. What will be the change in value as x changes from $0\cdot95$ to $1\cdot05$?

15. Find the equation of the tangent to the parabola $y = \frac{1}{4}x^2$ at $P(2k, k^2)$, and show that it meets the y-axis as far below the x-axis as P is above it.

16. What would you give as the best linear approximation to the function $f: x \to x^3-x$
  (a) over the domain $-1 \leqslant x \leqslant 1$;   (b) at the origin?

Note. A curve has a *point of inflexion* if its gradient changes from increasing to decreasing, or vice versa; at such a point the tangent crosses the curve.

10

INTEGRATION

1. AREA UNDER A CURVE

1.1 Area by estimation. We have long been familiar with the idea that the distance travelled by a body in a given time is represented by the area under its velocity-time graph. Here are a few examples.

(1) *Constant velocity*

Example 1. Suppose that a car travels for 5 minutes at a steady speed of 60 km/h. How far does it go? Obviously 5 km. It is convenient, however, to represent the result graphically. The velocity-time graph is a horizontal straight line. The area beneath the relevant part of the curve (Figure 1) is 5 square units, and one square unit represents $\frac{1 \text{ km}}{1 \text{ minute}} \times 1$ minute, that is, 1 km.

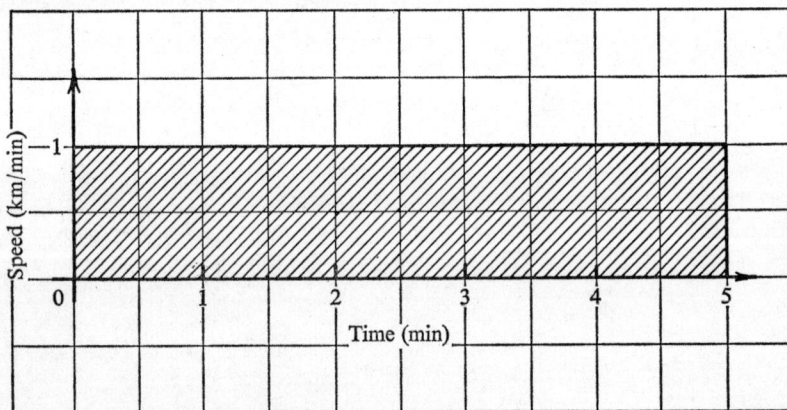

Fig. 1

(2) *Constant acceleration*

Example 2. Suppose that a stone falls under gravity increasing its velocity from 6 m per second to 28 m per second in 2 seconds. How far has it fallen in that time?

Figure 2 shows the velocity-time graph. In this case we could argue that, since the acceleration is constant, the average speed is 17 m/s and the

216

distance travelled is 34 m, which is again represented by the area under the velocity-time graph.

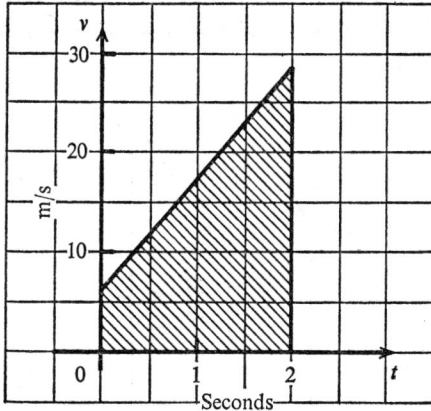

Fig. 2

We now consider the generalization of the above result in the case of variable acceleration.

(3) *Variable acceleration*

Example 3. The first stage of a rocket accelerates from rest to 75 m/s in 15 seconds in such a way that the velocity is given by

$$v = \tfrac{1}{3}t^2.$$

How far does it travel in that time?

The velocity-time graph is shown in Figure 3. We proceed to show that the distance is represented by the shaded area.

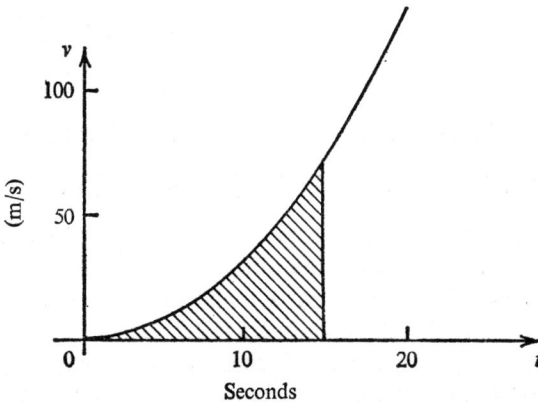

Fig. 3

At the moment we do not know a method for calculating this area exactly, or even if we can attach a meaning to such an 'area'. A method you may have used is to draw 'staircases' above and below the curve; then the areas of these 'staircases' give optimistic and pessimistic approximations, respectively.

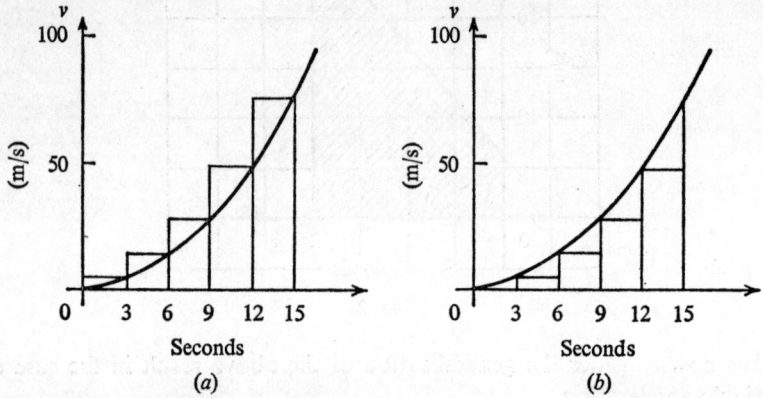

Fig. 4. (a) Optimistic approximation.   (b) Pessimistic approximation.

In this case taking steps of width 3 seconds (see Figure 4), we have limits for the area:

$$\text{optimistic approximation} = 3 \times \tfrac{9}{3} + 3 \times \tfrac{36}{3} + 3 \times \tfrac{81}{3} + 3 \times \tfrac{144}{3} + 3 \times \tfrac{225}{3}$$

$$= 495 \text{ m/s} \times \text{seconds} = 495 \text{ m},$$

$$\text{pessimistic approximation} = 3 \times \tfrac{0}{3} + 3 \times \tfrac{9}{3} + 3 \times \tfrac{36}{3} + 3 \times \tfrac{81}{3} + 3 \times \tfrac{144}{3}$$

$$= 270 \text{ m/s} \times \text{seconds} = 270 \text{ m}.$$

If we now take steps of 1 second (see Figure 5) the limits are:

$$\text{optimistic: } \tfrac{1}{3} \times 1^2 + \tfrac{1}{3} \times 2^2 + \tfrac{1}{3} \times 3^2 + \tfrac{1}{3} \times 4^2 + \ldots + \tfrac{1}{3} \times 15^2 = 413\tfrac{1}{3} \text{ m},$$

$$\text{pessimistic: } \tfrac{1}{3} \times 0^2 + \tfrac{1}{3} \times 1^2 + \tfrac{1}{3} \times 2^2 + \tfrac{1}{3} \times 3^2 + \ldots + \tfrac{1}{3} \times 14^2 = 338\tfrac{1}{3} \text{ m}.$$

As the steps decrease, the approximations get closer to each other. This can be seen in Figure 6 where the differences in each column have been shaded. Imagine these shaded parts to be moved across to occupy the rectangle $ABCD$. Their total area is $225 \times AB$. Hence as AB decreases this area decreases, and the approximations get closer together. By making the steps sufficiently small the approximations can be made to differ by as little as desired. The pessimistic approximation gives the distance which

218

would be covered if the velocity was increased in a series of jumps at the end of the time-intervals. Thus in Figure 4 the pessimistic approximation represents 3 seconds at 0 m/s, 3 seconds at 3 m/s, 3 seconds at 12 m/s and 3 seconds at 48 m/s; and this is clearly less than the actual distance travelled.

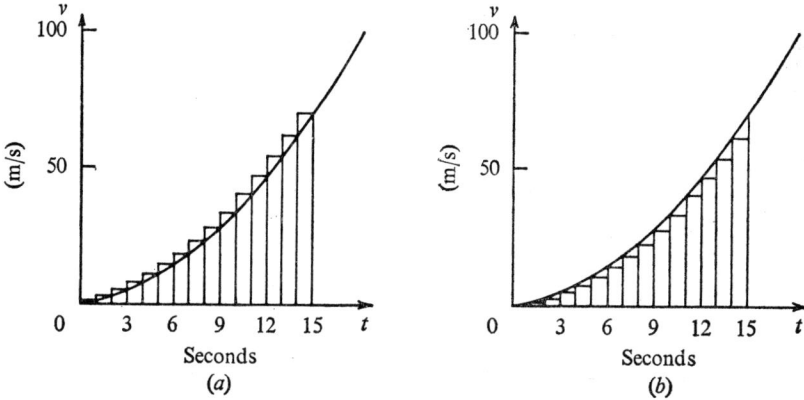

Fig. 5

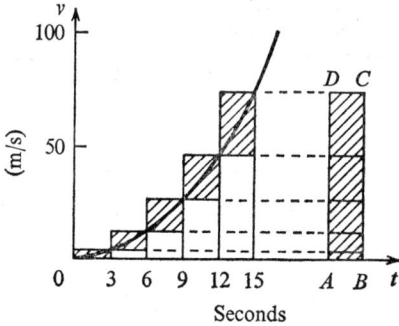

Fig. 6

Similarly the optimistic approximation gives the distance which would be covered if the velocity was increased in a series of jumps at the beginnings of the time-intervals, which is greater than the true distance. We conclude therefore that the actual distance travelled lies between the two approximations. But 'the area under the curve' also lies between them. Thus if we could show that each of the approximations tended to the same limit, we would have shown how to define 'the area under the curve' and that this was equal to the distance travelled. But, looking at Figure 6, we see that the difference between the two approximations is equal to the area of the rectangle *ABCD* and this can be made as small as we please by increasing the number of partitions. It can be proved that when this happens there is a real number equal to the common limit of the approximations; we define the area under the curve to be this number, and once again the area under the curve is equal to the distance gone.

1.2 The derivative of an area. The area under the velocity-time curve, then, represents the distance travelled; but the distance s and velocity v are also connected by the relation $v = ds/dt$. It must therefore be possible to regard the area as a function of t and to differentiate it. Nothing is gained by limiting our discussion to any particular form of velocity-time curve, so let us consider a completely general one; we shall assume only that it is continuous, that is, there are no sudden changes in velocity. We write $v = f(t)$.

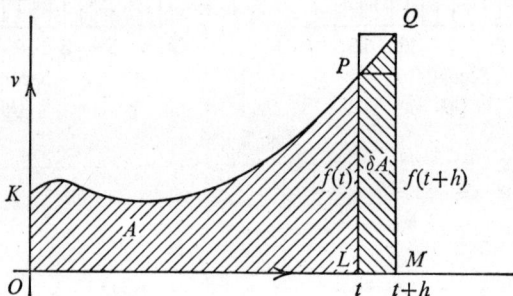

Fig. 7

Suppose A is the area under this curve from zero-hour up to time t (see Figure 7). Then A represents the distance described up to this time, and the velocity at time t should be given by

$$f(t) = v = \frac{dA}{dt}.$$

Let us look at this another way. There is a function $F: t \to A$ such that

$$F(t) = A = \text{shaded area } OKPL.$$

Now increase t to $t+h$; that is, consider a short time h following the instant represented by L. At the end of this time the area under the graph (or the distance covered) is
$$F(t+h) = \text{area } OKQM.$$

The distance covered in the interval h is represented by

$$\delta A = F(t+h) - F(t) = \text{area } LPQM.$$

This area is more than $LM \times LP$ (the pessimistic approximation) and less

220

than $LM \times MQ$ (the optimistic approximation)—or of course the other way round if MQ should happen to be less than LP.† In symbols

$$F(t+h) - F(t) \text{ lies between } hf(t) \text{ and } hf(t+h)$$
$$\Rightarrow \frac{F(t+h) - F(t)}{h} \text{ lies between } f(t) \text{ and } f(t+h).$$

But $F'(t)$ is the limit of this ratio

$$\frac{F(t+h) - F(t)}{h} \left(= \frac{\delta A}{\delta t} \right)$$

as h tends to zero; and as h tends to zero $f(t+h)$ tends to $f(t)$, since we have assumed that the velocity is continuous. Therefore

$$dA/dt = F'(t) = f(t).$$

We can, as it were, imagine a piston moving to the right under the curve and fitting it all the time. At time t the piston is at PL, and the rate at which the area $A = F(t)$ is growing is just equal to the height of the piston at this moment, that is, to $PL = f(t)$.

1.3 Calculation of area. We can therefore calculate A, given the shape of the function f, provided we can find a function F whose derived function is f. To illustrate the method, let us return to Example 3. We wish to find the area under the graph of $f: t \to \frac{1}{3}t^2$ from $t = 0$ to $t = 15$. Can we find a function F whose derived function is f? One such function is $F: t \to \frac{1}{9}t^3$; another is $F: t \to \frac{1}{9}t^3 + 1$. In fact, $F: t \to \frac{1}{9}t^3 + k$, where k is any constant, will have $\frac{1}{3}t^2$ for its derivative.

In our case only one of these functions will do, since $F(0) = 0$. This means that $k = 0$ and $F(t) = \frac{1}{9}t^3$. Hence $F(15) = \frac{1}{9} \times 15^3 = 375$ m, and this is the area required. The rocket travels 375 m in the first 15 seconds.

Example 4. What is the area under the graph of $y = x^2 + 2x + 3$ from $x = 1$ to 2? (Figure 8.)

The function shown by the graph is $f: x \to x^2 + 2x + 3$, and the area under this graph from some fixed left-hand end-point up to x is $A = F(x)$, where F is *some* function whose derived function is f. We can guess such an F to be

$$F: x \to \frac{1}{3}x^3 + x^2 + 3x + k$$

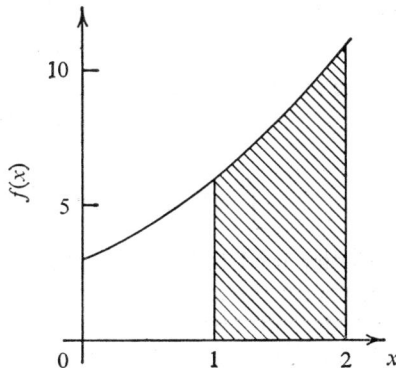
Fig. 8

† Strictly our argument only holds for functions f which are steadily increasing or steadily decreasing, or have a finite number of turning points, so that their domain can be divided up into parts in which one or other is the case.

221

(as is easily verified) for any constant k. The area we wish to find must be $F(2) - F(1)$, and this is

$$(\tfrac{1}{3}.8 + 4 + 6 + k) - (\tfrac{1}{3} + 1 + 3 + k) = 8\tfrac{1}{3}.$$

Exercise A

In Questions 1–3, draw a velocity-time graph for values of t from 1 to 3, estimate the area under it, and check by the method of Example 3. The units are metre-second units.

1. $v = t^2 + t$.   **2.** $v = t^3$.   **3.** $v = t^2 - 2t + 3$.

4. The velocity in m/s of a particle moving in a straight line is given by $v = 3t^2 + 2t$. Find how far it travels in the first 10 seconds.

5. The acceleration in m/s^2 of a bead moving on a straight wire is given by $a = 3t^2 + 2t$.
If it starts from rest at $t = 0$, finds it velocity after 2 seconds. How far has it then gone?

6. What does the area under an acceleration-time graph represent? If a particle, starting from rest, accelerates so that $a = t^2$, find the velocity after 4 seconds:
(a) by drawing a graph and estimating the area under it;
(b) by a method similar to that of Example 3.

7. What is $f(x)$ if $f'(x) = x^3 - x$ and $f(0) = 1$?

8. What is $f(x)$ if $f'(x) = 5x^4 - 3x^2 + 1$ and $f(2) = 3$?

1.4   Some language and notation. The process of finding a function F, given f, such that $F' = f$ is called anti-differentiation or *integration*, and any function F so found is called *an integral* of f. We may write

$$F = \textstyle\int f \quad \text{(read 'an integral of } f\text{')},$$

but the more usual notation is

$$F(x) = \int^x f(u)\, du,$$

or $$\int^x f(t)\, dt \quad \text{(often, loosely, written } \textstyle\int f(x)\, dx\text{)}.$$

The letter used (u or t) for the typical element mapped by f is of no importance; the area under the graph of $f(t)$, with a fixed left-hand end-point, depends only on the position of its right-hand end, here called x.

As in the case of differentiation, there is a historic reason for the notation. If we look back at Example 3, we see that the 'pessimistic approximation' is the sum of a set of rectangles, of which a typical one has area $f(t)\,\delta t$, where δt ('a small bit of t') is the time interval involved. In symbols

$$A \simeq \Sigma f(t)\, \delta t,$$

where Σ stands for 'the sum of'; when we take the limit of this approximation we write $A = \int f(t)\,dt$ for the exact result.

The 'integral sign' $\int$ is simply a long S, meaning 'the limiting sum of'. When we take the area from $t = 1$ to 2, we can indicate this in the notation by writing $\int_1^2 f(t)\,dt$ (read 'the integral from 1 to 2 of $f(t)$ with respect to t'), and this is, as we have seen, $F(2) - F(1)$. To summarize: if $F' = f$, then

$$\int^x f(t)\,dt = F(x);$$

and, in general,
$$\int_a^b f(t)\,dt = F(b) - F(a).$$

Example 5. Suppose we know that the acceleration of a particle is given by the formula $a = 6t - 6$, and that when $t = 0$, $v = 2$ and $s = 0$. Since $dv/dt = a$, we need to find a function f so that $v = f(t)$ and $f'(t) = 6t - 6$. Intelligent guessing might give $f(t) = 3t^2 - 6t$. However, in this case $f(0) = 2$. Referring to Figure 9 we see that $f(t) = 3t^2 - 6t + c$ would serve, where c is any constant.

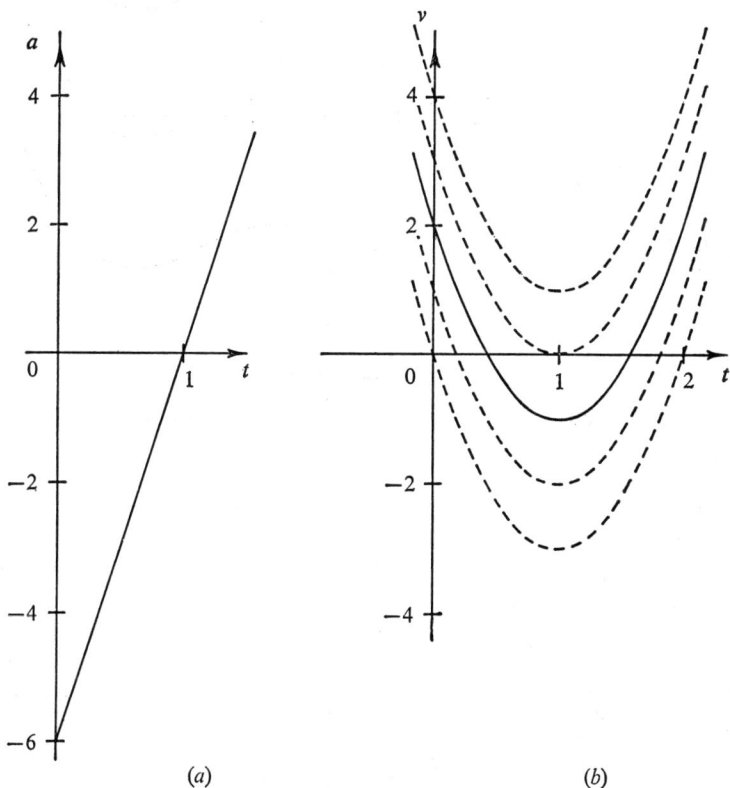

(a)   (b)

Fig. 9

If $f(t) = 3t^2 - 6t + c$ then $f(0) = c$, and thus the appropriate constant
is 2. Further,

$$\frac{ds}{dt} = V$$

$$\Rightarrow \frac{ds}{dt} = 3t^2 - 6t + 2$$

$$\Rightarrow s = t^3 - 3t^2 + 2t + d.$$

Since in this case $s = 0$ when $t = 0$, we have $d = 0$ and

$$s = t^3 - 3t^2 + 2t.$$

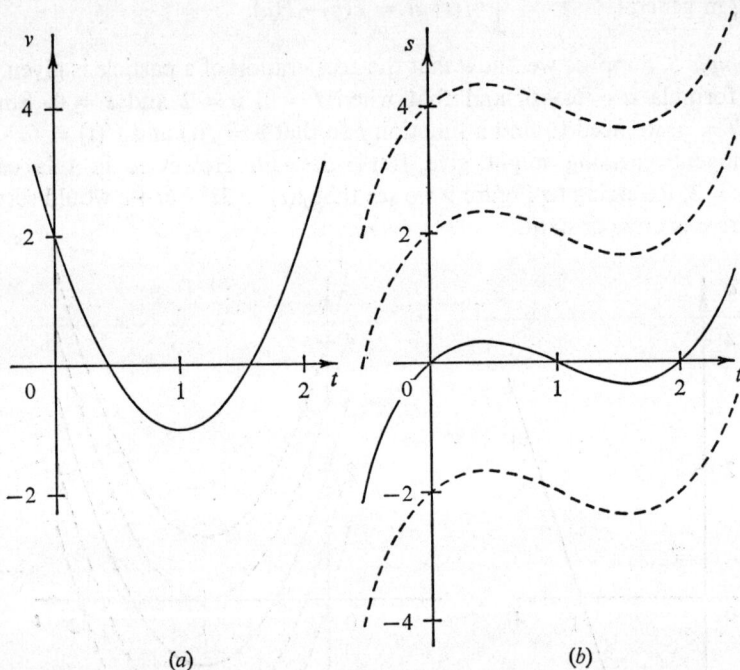

(a)   (b)

Fig. 10

It is interesting to examine the graphs. In Figure 9 any of the velocity
curves will give the same acceleration, but only one has $v = 2$ at zero time.
Similarly, in Figure 10, the velocity curve could represent the gradient
function of any one of the infinite 'family' of distance functions

$$s = t^3 - 3t^2 + 2 + d,$$

but only one of these has $s = 0$ at $t = 0$.

Example 6. Evaluate and interpret

$$\int_2^5 (x^2 - x) \, dx.$$

224

If $\qquad\qquad F'(x) = x^2 - x, \quad F(x) = \frac{1}{3}x^3 - \frac{1}{2}x^2 + k.$

Then $\qquad\qquad \displaystyle\int_2^5 (x^2 - x)\, dx = F(5) - F(2)$

$$= \tfrac{125}{3} - \tfrac{25}{2} + k - (\tfrac{8}{3} - 2 + k)$$

$$= 28\tfrac{1}{2}.$$

This is the area under the graph of $y = x^2 - x$ from $x = 2$ to $x = 5$ (Figure 11).

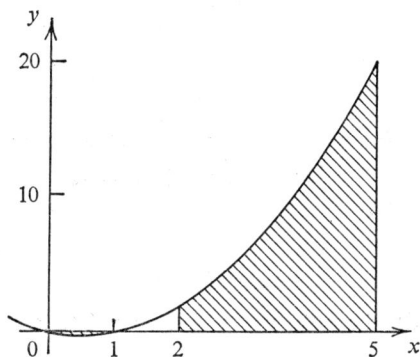

Fig. 11

Example 7. Evaluate and interpret

$$\int_0^1 (x^2 - x)\, dx.$$

With the same F as before,

$$\int_0^1 (x^2 - x)\, dx = F(1) - F(0)$$

$$= (\tfrac{1}{3} - \tfrac{1}{2} + k) - k$$

$$= -\tfrac{1}{6}.$$

This is the area 'under' the same graph from $x = 0$ to 1. We see that, to be consistent, when the graph is below the axis we must count the area as negative.

1.5   Some formal results. So far we have guessed the value of the function F. Basically, this is what integration involves, but guesswork can be guided and intelligent, and should always be checked by differentiation. For powers of x, however, the integral can be written down by rule, as the following table shows.

$g(x)$	$g'(x)$	$f(x)$	$\int f(x)\,dx$
k	0	0	k
x	1	1	$x+k$
x^2	$2x$	x	$\frac{1}{2}x^2+k$
x^3	$3x^2$	x^2	
x^4			

....

x^n	$n.x^{n-1}$	x^{n-1}	
x^{n+1}		x^n	

In each case $f(x)$ is obtained from $g'(x)$ by dividing by a number, and $\int f(x)\,dx$ (apart from the added constant k) is obtained from $g(x)$ by dividing by the same number. Complete the table yourself.

It may also be proved that, apart from the arbitrary constants k,

$$\int f(x)\,dx + \int g(x)\,dx = \int [f(x)+g(x)]\,dx$$
$$\int cf(x)\,dx = c\int f(x)\,dx,$$

where c is any number.

Exercise B

1. Find the area under the curve $y = x^2$ from $x = 0$ to $x = 2$.

2. Find the area under the curve $y = \frac{1}{4}x^3$ from $x = 1$ to $x = 3$.

3. Find the area under the curve $y = x+\frac{1}{2}x^3$:
(a) from $x = 0$ to $x = 5$,
(b) from $x = 0$ to $x = 3$ and
(c) from $x = 3$ to $x = 5$.

$\frac{1}{2}x^2 + \frac{1}{8}x^4$

4. Evaluate the following:

(a) $\int_0^2 x^2\,dx$; (b) $\int_{-1}^1 (2x+3)\,dx$; (c) $\int_{-1}^1 (1-x^2)\,dx$;

(d) $\int_2^5 (x^3-3x^2+1)\,dx$; (e) $\int_0^2 x(x-2)\,dx$; (f) $\int_{-1}^1 x(x^2-1)\,dx$.

In each case sketch the graph and explain any surprising results.

5. A rocket accelerates from rest in two stages: first such that

$$a = \frac{3t^2}{10} \quad \text{for} \quad 0 \leqslant t \leqslant 10$$

and then such that
$$a = 3t+20 \quad \text{for} \quad 10 \leqslant t \leqslant 150.$$

Find by integration its velocity when $t = 150$. (See Example 9.)

6. A projectile is fired vertically upwards so that its velocity in m/s after t seconds is
$$v = 420-10t.$$
Find:
(a) the time to its greatest height;
(b) its greatest height.

226

7. A stone is thrown vertically downwards from the top of a tower with velocity 3 m/s. Its acceleration is 9·8 m/s² (the acceleration due to gravity). Find the velocity with which it strikes the ground 5 seconds later. How high is the tower?

8. An aircraft fires a projectile upwards from a height of 1300 m with a velocity of projection of 210 m/s. Assuming the acceleration due to gravity is 9·8 m/s², and neglecting the horizontal motion of the projectile, find:
  (a) the velocity of the projectile when $t = 10$ seconds;
  (b) the time taken to reach the greatest height;
  (c) the greatest height;
  (d) the time of striking the earth.

9. If a particle moves in a straight line so that its velocity after t seconds is

$$v(t) = 3t^2 - 6t + 2,$$

it being at the origin when $t = 0$, show that it passes through the origin twice more. Find the velocity and acceleration on these occasions. Draw graphs of the motion.

2. CONSTANT ACCELERATION

We know that $v = ds/dt$ and $a = dv/dt$ (where a, v, s, t stand respectively for acceleration, velocity, distance and time).

Now if the acceleration a is *constant*, we can integrate the equation $dv/dt = a$ and obtain

$$v = at + c. \tag{1}$$

In this, we normally write $c = u$, where u is the initial velocity; that is, the velocity at $t = 0$. Thus (1) becomes

$$v = u + at. \tag{2}$$

Integrating this we obtain (having written ds/dt for v)

$$s = ut + \tfrac{1}{2}at^2. \tag{3}$$

(Here the constant of integration is zero if we measure distance from the starting point.) Combining (2) and (3) we have

$$s = t(u + \tfrac{1}{2}at)$$
$$= \tfrac{1}{2}t(2u + at)$$
$$= \tfrac{1}{2}t(u + v). \tag{4}$$

Since $\qquad\qquad v - u = at$ from (2),

and $\qquad\qquad v + u = 2s/t$ from (4),

we can multiply these together to obtain

$$v^2 - u^2 = 2as. \tag{5}$$

Collecting these results together we have

$$\begin{cases} v = u + at, \\ s = ut + \tfrac{1}{2}at^2, \\ s = \tfrac{1}{2}(u+v)t, \\ v^2 = u^2 + 2as. \end{cases}$$

Note that we cannot integrate $ds/dt = v$ to obtain $s = vt + c$. (Why not?)

Example 8. A car starts from rest with constant acceleration 1 m/s² for 20 seconds; it then moves with a reduced acceleration of $\tfrac{2}{3}$ m/s² for the next 10 seconds, and finally travels at a steady speed of $26\tfrac{2}{3}$ m/s. How long does the car take to travel 500 m?

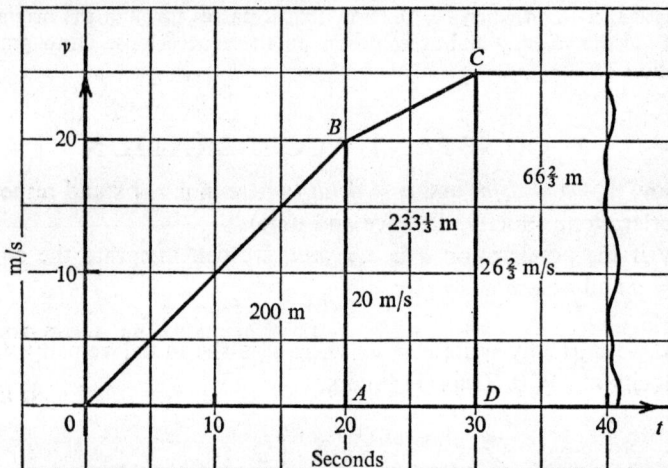

Fig. 12

This problem is best solved directly from the velocity-time graph (Figure 12). The area of the triangle OAB is $\tfrac{1}{2} \times 20$ m/s $\times 20$ s $= 200$ m. The area of the trapezium $ABCD$ is $\tfrac{1}{2} \times (20 + 26\tfrac{2}{3})$ m/s $\times 10$ s $= 233\tfrac{1}{3}$ m. There remains $66\tfrac{2}{3}$ m of rectangular area on the right; the height of this rectangle is $26\tfrac{2}{3}$ m/s, so that its base must be $2\tfrac{1}{2}$ s. The total time is therefore $32\tfrac{1}{2}$ s.

Example 9. The acceleration a m/s² of a car t seconds after starting from rest is given by $a = t^3/125$ for $0 \leqslant t \leqslant 5$ and by $a = t$ for $t > 5$.

What is the car's speed after 6 seconds?

Clearly for the first 5 seconds

$$v = \frac{t^4}{500}$$

and thereafter

$$v = \tfrac{1}{2}t^2 + c.$$

228

But how do we calculate c in this case? We must assume the obvious physical fact that the velocity is continuous even when the acceleration changes. Thus $v(5) = \frac{625}{500} = 1\frac{1}{4}$ and so $c = 1\frac{1}{4} - \frac{1}{2}.25 = -11\frac{1}{4}$. Hence $v(6) = \frac{1}{2}.36 - 11\frac{1}{4} = 6\frac{3}{4}$ m/s. The car's speed is $6\frac{3}{4}$ m/s.

Exercise C

1. A car accelerates uniformly from 6 m/s to 10 m/s in 15 seconds. What is its acceleration? How far does it go in the time?

2. A stone is dropped from rest and falls freely for 12 seconds with an acceleration of 9·8 m/s². How far does it drop?

3. If the stone in Question 2 had been dropped from a lift travelling upwards at 3 m/s find how far it would drop. How long does it take to fall 2100 m?

4. Answer the first part of Question 3 if, owing to resistances, the velocity never exceeds 28 m/s (assume that it still accelerates at 9·8 m/s²).

5. A lorry travelling at 13 m/s overtakes a car travelling at 10 m/s. 30 seconds later the car accelerates at $\frac{2}{9}$ m/s² up to a steady speed of 20 m/s. After how long does the car overtake the lorry and how far does it go in the time?

6. The acceleration of a car starting from rest is given, after t seconds, by $a = \frac{2}{3}t^3$ for the first 4 seconds. The car then accelerates at 2 m/s² for 15 seconds and then at $\frac{3}{4}$ m/s² for 12 seconds. What is its final speed? How far does it go?

3. OTHER USES OF INTEGRATION

Example 10. Find the volume of a conical heap of sand whose base area is A m² and whose height is p m (see Figure 13).

There is no need for the base region to be a circle, provided only that

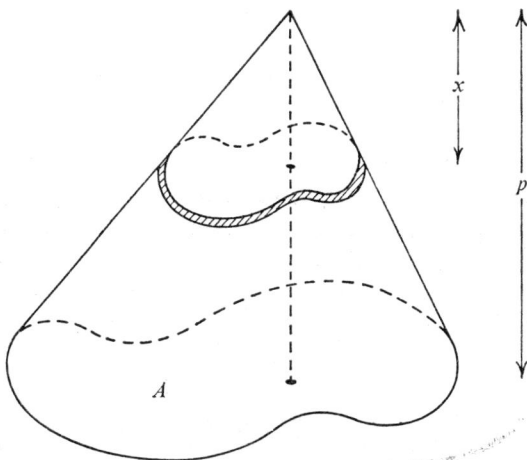

Fig. 13

the cone is generated by straight lines joining its vertex to the perimeter of the base.

There is a definite volume of sand lying above a horizontal plane at a depth x m below the vertex; there is therefore a function V mapping x onto this volume, which is then $V(x)$ m³. Between the planes at depth x m and $(x+h)$ m there is a slice of volume $V(x+h)-V(x)$ m³. The upper surface of this slice is a region similar to the base region, of area Ax^2/p^2 m²; the lower surface has area $A(x+h)^2/p^2$ m². It is intuitively obvious that the volume of the slice lies between

$$\frac{hAx^2}{p^2}\text{ m}^3 \quad\text{and}\quad \frac{hA(x+h)^2}{p^2}\text{ m}^3;$$

Hence
$$\frac{Ax^2}{p^2} < \frac{V(x+h)-V(x)}{h} < \frac{A(x+h)^2}{p^2},$$

and in the limit we must have

$$V'(x) = \frac{Ax^2}{p^2}.$$

Since $V(0) = 0$, we have at once

$$V(x) = \frac{Ax^3}{3p^2}$$

and the volume of the whole cone,

$$V(p)\text{ m}^3 = \frac{Ap}{3}\text{ m}^3.$$

The situation occurring in this example is typical. We break up the quantity we are trying to calculate into 'slices' depending on a variable quantity x; we show that the contribution of each slice is between $f(x).h$ and $f(x+h).h$, where h is the change in x across the slice. Then the total of all these contributions tends to

$$\int_a^b f(x)\ dx,$$

where a and b are the extreme values of x.

Example 11. Water escapes through a leak in a dam in such a way that t minutes after the leak is sprung the rate of leakage is $3+\frac{1}{2}t^2$ l/m. How long will it be before 1000 litres have leaked away?

From t to $t+h$ minutes after zero, between $(3+\frac{1}{2}t^2)h$ and $[3+\frac{1}{2}(t+h)^2]h$ litres of water pass through the leak. This is the typical integration situation; if the volume escaped up to time t is $V(t)$, then

$$V'(t) = 3+\tfrac{1}{2}t^2;$$
$$\Rightarrow V(t) = \int(3+\tfrac{1}{2}t^2)dt$$
$$= 3t+\tfrac{1}{6}t^3+c.$$

But $V(0) = 0$, so that $c = 0$, and

$$V(t) = 3t+\tfrac{1}{6}t^3.$$

230

To find when this is equal to 1000, we must solve

$$3t + \tfrac{1}{6}t^3 = 1000$$

$$\Leftrightarrow t^3 + 18t = 6000.$$

When $t = 17$, $t^3 + 18t = 4913 + 306 = 5219$;

when $t = 18$, $t^3 + 18t = 5832 + 324 = 6156.$

It takes nearly 18 minutes.

Example 12. When a spring of natural length 20 cm is pulled out an extra x cm, the tension in it is $300x$ N. How much work is done in extending it to 30 cm? [The work done by a constant force is the force multiplied by the distance it moves the point it is applied to in its direction.]

The work done in stretching the spring from $20 + x$ cm to $20 + x + h$ cm lies between $h.\,3x$ and $h.\,3(x+h)$ Nm. Again we wish to sum these contributions from $x = 0$ to $x = 10$—a typical integration situation. We conclude that the total work is

$$\int_0^{10} 3x\,dx = 1\!\cdot\!5x^2 \quad \text{when} \quad x = 10$$

$$= 150 \text{ Nm.}$$

To justify this, we should say that there is a function W of x so that the work done in extending the spring by x cm is $W(x)$; then

$$h.\,3x < W(x+h) - W(x) < h.\,3(x+h)$$

$$\Rightarrow 3x < \frac{W(x+h) - W(x)}{h} < 3(x+h)$$

$$\Rightarrow W'(x) = 3x,$$

and so on. But once we recognize the situation we can write down the integral at once.

Exercise D

1. The part of the curve $y = x^2$ from $x = 0$ to $x = 5$ is revolved about the x-axis to form a *solid of revolution.*

Show that the part of its volume from x to $x+h$ lies between $\pi x^4.\,h$ and $\pi(x+h)^4.\,h$. Use the method of Example 10 to find the total volume.

2. If the curve in Question 1 had been the graph of $y = f(x)$, show that the volume would have been

$$\pi \int_0^5 [f(x)]^2 \, dx.$$

Evaluate this when $f(x) = 3 + 5x - x^2$ and illustrate with a sketch of the solid.

3. Find the volume of the *paraboloid* formed by rotating the curve in Question 1 about the *y-axis.*

4. Find the volume of a sphere of radius a by first showing that a slice between parallel planes at distances x and $x+h$ from the centre has volume approximately $\pi h(a^2-x^2)$.

5. If the surface area of a sphere of radius r is $S(r)$, explain why the rate of change of its volume with r is $S(r)$. Hence find $S(a)$.

6. A barrel has the form of a solid formed by rotating the graph of

$$f: x \to 1 + \frac{x}{4} - \frac{3x^2}{64} \quad (0 \leqslant x \leqslant 4)$$

about the x-axis, the units being metres. Sketch the shape of the barrel and find its volume.

7. If it requires a force of $20x$ N to stretch a piece of elastic of length 24 cm by x cm, find the work done in stretching it from 48 cm to 72 cm.

8. A railway buffer requires a force of $1250x^2$ N to compress it x m. How much work is needed to compress it from 0·25 m to 0·50 m?

9. A forest fire is spreading at a rate of $3t^2+4t^3$ m²/h. What area is burnt out in the first 3 h? In the fourth hour?

10. In a small town the density of population at x km from the centre is about $1500x$ persons per km² when $0 \leqslant x \leqslant 1$, and about $250(3x-x^2)$ persons per km² for $1 < x \leqslant 3$; it is zero for $x > 3$. Find the population of the town.

11. A crowd is gathering round a public orator at a rate of $100(5x-3x^2)$ persons per hour, x hours after he starts to speak.
  (a) What is the largest crowd he has?
  (b) After how long is this?
  (c) When does his audience completely disappear?

12. Calculate the volume of a vase 20 cm high which has a radius of

$$6-x+x^2/10 \text{ cm}$$

at a height of x cm. Sketch the vase.

Miscellaneous Exercise (on Differentiation and Integration)

1. Define the terms speed and acceleration. What distinguishes the shape of a velocity-time graph at a point of:
  (a) zero acceleration;
  (b) discontinuous acceleration?

2. Find from first principles the gradient of the functions
  (a) $f: x \to x^3-3x$;   (b) $f: x \to 4/x$.

3. A particle moves in a straight line so that after time t seconds its velocity is v m/s, where $v = 4t-t^2$. Find its distance-time equation and its acceleration at any time.
  When in the first five seconds is it furthest from the origin?
  When in the first twenty?

4. If $y = x+1/x^2$, find dy/dx and sketch the graph of y.

5. A cylindrical jar, with a base but no lid, is to be made from 6π cm² of metal. What is its greatest possible volume?

232

6. A cylindrical jar, with base and lid, is to have a volume of 54π cm³. What is the least amount of metal from which it can be made?

7. Use linear approximation to estimate $(6\cdot003)^6$ and $\sqrt[3]{(7\cdot8)}$.

8. A particle moving in a straight line has a velocity given by the formula $v = 4/(1+t^2)$. By considering the optimistic and pessimistic approximations at intervals of $\frac{1}{10}$ second and taking their average, estimate how far it goes in the first second.

9. For a particle moving in a straight line with *constant* acceleration, prove, in the usual notation, that
$$s = vt - \tfrac{1}{2}at^2.$$

10. A car travelling at 20 m/s is overtaken by a car travelling at 30 m/s. The faster car is accelerating at $\frac{1}{2}$ m/s², and continues to do so for 10 seconds and thereafter maintains a speed of 35 m/s. 30 seconds after being overtaken, the other car accelerates from 20 m/s to 40 m/s in 20 seconds, and then maintains that speed. After how long does the first car catch up on the second one?

11. Find the area under the curve $y = 1/x^2$ from $x = 2$ to $x = 4$
    (*a*) by integration;
    (*b*) by taking the average of the optimistic and pessimistic approximations of 20 strips of width $1/10$.

12. A car accelerates from rest, its acceleration a m/s² being given by the formula
$$a = \tfrac{2}{3}t^2 + 1,$$
where t seconds is the time after starting. After 3 seconds the acceleration changes to a steady 2 m/s²; and after a further 2 seconds the car ceases to accelerate. Find formulae for its speed at various times and find how far it goes in the first 10 seconds.

13. The curve $y = 1 + x^2$ from $x = 0$ to $x = 2$ is rotated about the x-axis through 360°. Find the volume of the solid of revolution so formed.

*14. A wooden napkin ring is in the form of a solid sphere of radius 2·5 cm with a cylindrical hole of radius 2 cm drilled out of it. Find the volume of wood in the ring.

*15. A pyramidal heap of soil is compressed at the bottom in such a way that at depth x m below the apex it weights $80 + x/8$ kg/m³. The heap is $1\frac{1}{3}$ m high and rests on a base $1\frac{2}{3}$ m square. Find its total mass.

*16. Find the area of the region between the curves $y = 1 + x^2$ and $y = 1 + \frac{1}{2}x^3$ and the volume of the ring formed by rotating this region about the axis of x.

11
CIRCULAR MEASURE AND CIRCULAR FUNCTIONS

1. THE SINE WAVE

There are many situations which occur in everyday life where some physical quantity fluctuates in a regular manner, repeating its changes periodically at equal intervals of time. In many of these the variation, plotted against the time, approximates more or less closely to the form of the sine curve, in which case it is said to be *sinusoidal*. We give a number of examples.

1.1 Physical examples.

1. *The Big Wheel.* The Big Wheel at a fairground turns at a constant speed, making one complete revolution in 10 seconds.

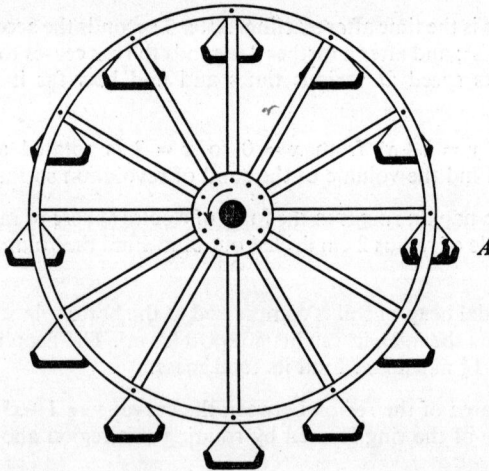

Fig. 1

How does the height of a passenger above the centre of the wheel vary?

Figure 2 shows the height, measured from position A, plotted against time. This is of course the familiar sine curve itself, the height of the passenger above A being the 'sideways displacement' $r \sin \theta$, where θ is the angle turned through by the wheel from the position where the passenger is at A, and r is the radius of the wheel.

234

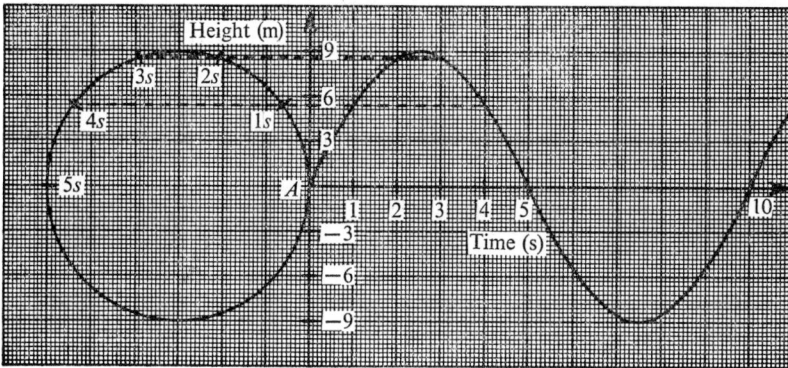

Fig. 2

2. *The pendulum.* If a pendulum swings through a total angle α on each side of the vertical, the angle θ it makes with the vertical at any time varies with time as shown in Figure 3. Careful measurement will show that this is not an exact sine curve, but the difference is very slight.

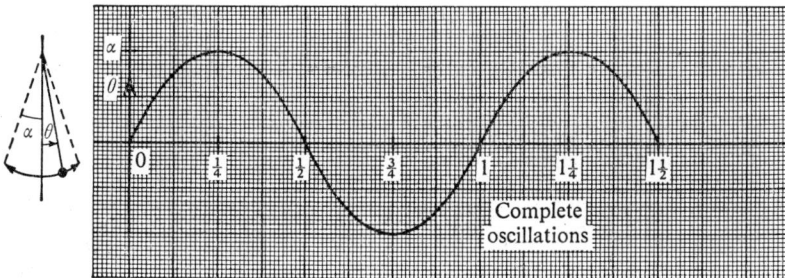

Fig. 3

3. *Sound vibrations.* If a tuning fork is struck, the distance between its prongs also varies with the time in the same general way; so does the pressure of the air at a given point of an organ pipe, or the displacement of a definite point on a vibrating violin string. All these show periodic movements which under suitable circumstances can be represented by a sine wave.

4. *Alternating current.* When a coil of wire rotates in a magnetic field (see Figure 4) a current is produced in the coil, which is not constant but varies with the position of the coil as shown in Figure 5. This is *alternating current,* such as is supplied in ordinary domestic electricity, and its graph is once again a sine curve. Each complete oscillation of current, from zero to its positive peak, back to zero and down to its negative peak, and back to zero again, is called a *cycle*.

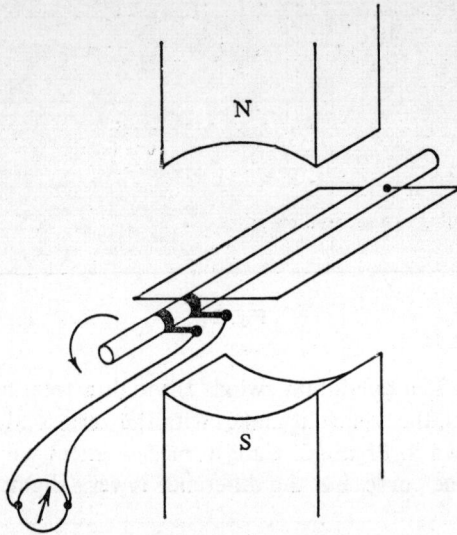

Fig. 4

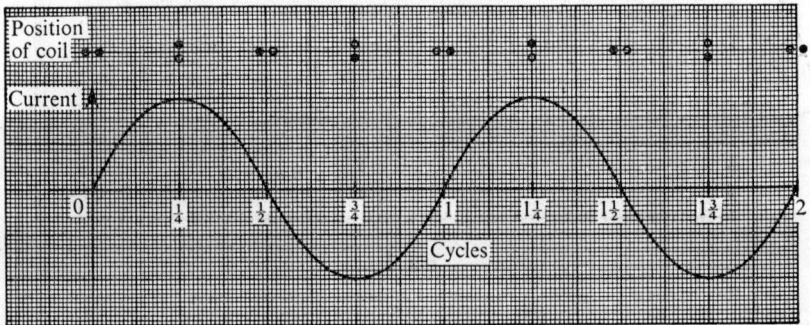

Fig. 5

Exercise A

Use the graph of Figure 2 in answering Questions 1–4.

1. The Big Wheel has radius 9 m (see Example 1). After what time does the passenger
    (*a*) first return to position *A*;
    (*b*) first reach $4\frac{1}{2}$ m above the level of *A*;
    (*c*) next reach $4\frac{1}{2}$ m above the level of *A*?

236

2. A pendulum swings out to an angle of 15°, and its period (that is, the time taken to swing from one extreme to the other, and back) is 3 seconds. What is the time between

   (a) 2 consecutive vertical positions;

   (b) 2 consecutive positions at 10° to the vertical?

3. The frequency of a simple alternating current is 50 cycles per second. What is its period?

4. At a certain port, the tide has a greatest height of 2 m above the mean tide level at 7 a.m. and its period is 12 hours. Assuming the motion produces a simple sine wave, find the height of the tide at 12 noon.

5. How would you obtain the graph of

$$x \to \cos x°$$

from the graph of $x \to \sin x°$?

Write in a simpler form the functions

   (a) $x \to \sin (x+90)°$;   (b) $x \to \cos (x-90)°$.

6. How would you obtain the graphs of the following functions from that of $x \to \sin x°$?

   (a) $x \to \sin 2x°$;   (b) $x \to \sin 3x°$;

   (c) $x \to \sin \frac{1}{2}x°$;   (d) $x \to \cos 2x°$.

7. Without accurate calculation give a rough sketch of the gradient function of $x \to \sin x°$.

8. Give rough sketches of the graphs of

   (a) $x \to \sin (x+30)°$;   (b) $x \to \sin (x+120)°$;

   (c) $x \to \cos (x+45)°$;   (d) $x \to \cos (x+135)°$.

2. ANGLE MEASURE

2.1   The graph of the sine function. If x is a real number, then we can define a function $f: x \to \sin x°$ whose range will be the real numbers from -1 to $+1$. The domain of f is the whole of the real numbers. Figure 6 shows part of the graph of f.

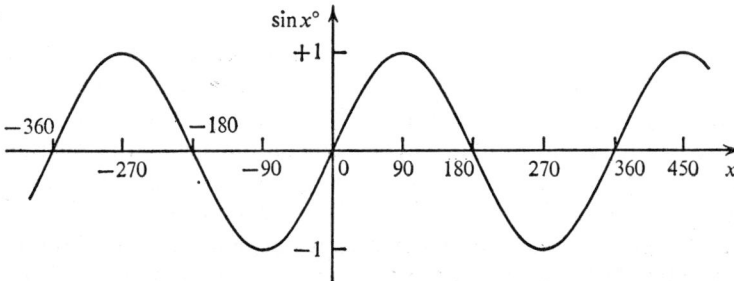

Fig. 6

This function is oscillatory and periodic and, as we have seen, it is a model of the many physical phenomena which behave in this way. For this purpose it has, however, one serious disadvantage—the great disparity of the scales on the two axes. The maximum gradient of the function, when $x = 0$, is only about 1/60. The following exercise is meant to suggest how best we could alter this.

Exercise B

1. Suppose the line $y = x$ is drawn on graph paper. Without altering the line, let us alter the units on the x-axis so that the new unit is ten times as long as the old; i.e. the division formerly marked 10 is now marked 1. The units on the y-axis are unaltered. What is now the gradient of the line?

2. Consider the function $x \to \sin x_h$, where x_h is the measure of an angle in units of a half-turn (180°). Will its graph look like that of Figure 5? What will be the graduation on the x-axis where 90 stood before?

3. If the gradient of $\sin x°$ at $x = 0$ is 1/60, what will be the gradient of $\sin x_h$?

4. If the unit of angle is a whole-turn (360°) and x_t is the measure of an angle in whole-turns, what is the gradient of $\sin x_t$ at $x = 0$, approximately? Sketch the graph of the gradient of $\sin x_t$ from $x_t = 0$ to 1.

5. If we chose a unit of angle to make the gradient of $\sin x$ unity at $x = 0$, when x is measured in these units, about how many degrees would the unit be?

6. The gradient of $\sin x°$ can be found approximately from ordinary trigonometric tables, by considering the change in $\sin x°$ when x changes by 1°. Copy and complete the following table:

x	0	1	30	31	45	46	60	61	89	90
$\sin x°$										
Average gradient										
$\cos x°$										

What do your results suggest?

2.2 Circular measure. In all the physical examples of sine waves that we have considered, it should be obvious that the rate of change of the physical quantity with the time is likely to be important. The previous exercise will have shown that this depends on our method of measuring angles. If we measure them in degrees, the gradient of $\sin x°$ at the origin turns out to be about 0·0175, which is an awkward and inconveniently small number. It arises because the ancient Babylonians decided to divide the circle into 360 equal parts. Modern mathematicians prefer to eliminate the Babylonians and to start again with a fresh approach.

Let us decide to choose a unit of angle measure so that the gradient of sin x, where x is measured in our new units, will have the value 1 at $x = 0$. What will the unit be? To answer this, consider Figure 7. In this figure, the radius OU is of length 1 unit; the length of $OM(y)$ is therefore sin θ.

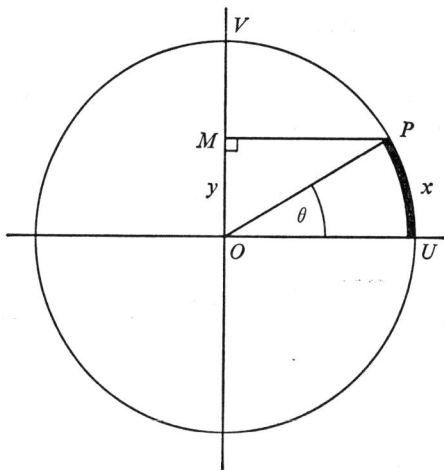

Fig. 7

The gradient of sin x when $x = 0$ is the limit of the ratio

$$\frac{\sin x}{x} \quad \text{as} \quad x \to 0.$$

We write this
$$\lim_{x \to 0} \frac{\sin x}{x}.$$

To make this limit 1, we must choose x, proportional to θ, so that $y/x \to 1$ as $\theta \to 0$. The obvious choice is to make x equal to the length of the arc UP. This is certainly proportional to θ, and y $(= OM)$ and x $(= UP)$ are more and more nearly equal as θ becomes small. This means that the unit of angle is that angle for which $UP = 1$; i.e. the angle which subtends at the centre O an arc equal in length to the radius of the circle. This angle is called a RADIAN.

Measure the diameter of a circular cylinder (a tin or tin lid will do). Roll the cylinder a distance equal to its radius along a level surface. Through what angle (in degrees) has it turned? (See Figure 8.)

Repeat for different sized cylinders, in each case rolling them distances equal to their radii. What result do you get? Did you expect this result?

In Figure 9, C_1 is a circle of radius 1 unit, and DE is of length 1 unit. C_2 is a concentric circle of radius r units.

By similarity, AB is of length r units, that is, AB is equal to the radius of C_2.

239

Thus, an arc of a circle of length equal to the radius always subtends the same angle at the centre, the angle we have called 1 radian.

Now the circumference of a circle of radius r is $2\pi r$, and each length r of the circumference subtends an angle of 1 radian at the centre. Hence,

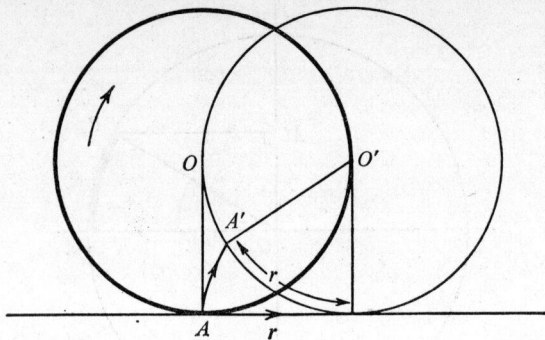

Fig. 8

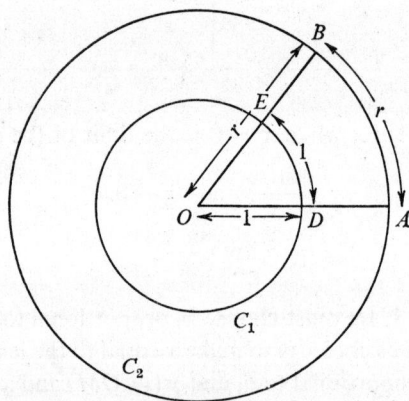

Fig. 9

the whole circumference subtends an angle of 2π radians at the centre. But this angle is 360°. Hence

$$2\pi \text{ radians} = 360°$$

$$\Rightarrow 1 \text{ radian} = \frac{360°}{2\pi} \simeq 57° \, 18'.$$

Conversion from degrees to radians and vice versa can be effected by the use of this equation; or can be read out of conversion tables.

The measures of such familiar angles as half-turns, quarter-turns, and so on are usually quoted not as decimals, but as fractions of π radians; if we want to draw particular attention to the fact that the object we are measuring is an angle and the unit is a radian, we write the unit as 1^c, and a quarter-turn, for example, is $\frac{1}{2}\pi^c$; but in many physical applications we are not talking about an angle but simply about a number, and the ^{c} is usually left out. For the present, $\sin x$ can be taken to mean $\sin x^c$.

Important equivalents are:

$$30° = \tfrac{1}{6}\pi \text{ radians},$$
$$45° = \tfrac{1}{4}\pi \text{ radians},$$
$$60° = \tfrac{1}{3}\pi \text{ radians},$$
$$90° = \tfrac{1}{2}\pi \text{ radians},$$
$$120° = \tfrac{2}{3}\pi \text{ radians},$$
$$180° = \pi \text{ radians}.$$

Exercise C

1. Write down the measures in radians (using π) of: 135°, 270°, 300°, 390°, 720°. 765°, 67½°, 36°, 108°, 110°.

2. (*a*) Express in degrees: $0\cdot94^c$, $2\cdot04^c$, $4\cdot91^c$.
   (*b*) Express in radians: 42°, 198°, 427·8°.

3. Write down the measures in degrees of the angles whose radian measures are:
$$\tfrac{1}{8}\pi, \ \tfrac{3}{10}\pi, \ \tfrac{7}{2}\pi, \ \tfrac{5}{6}\pi, \ \tfrac{11}{12}\pi, \ \tfrac{7}{5}\pi, \ 3\pi, \ \tfrac{9}{4}\pi, \ \tfrac{13}{18}\pi, \ \tfrac{27}{20}\pi.$$

4. What are $\sin \tfrac{1}{2}\pi$, $\cos \tfrac{1}{3}\pi$, $\tan \tfrac{3}{4}\pi$, $\cos \tfrac{5}{6}\pi$, $\sin \pi$, $\cos \tfrac{3}{2}\pi$, $\tan \tfrac{2}{3}\pi$, $\cos 2\pi$?

5. Simplify $\sin (\tfrac{1}{2}\pi - \theta)$, $\cos (\pi + \theta)$, $\tan (2\pi - \theta)$.

6. Show that for all x, $\cos (2\pi + x) = \cos x$ and state the corresponding results for $\sin (2\pi + x)$ and $\tan (\pi + x)$.

7. The radius of a bicycle wheel is 30 cm. What angle in radians does it turn through in travelling 60 cm?

8. Show that the length s of an arc of a circle of radius r subtending an angle of θ radians at the centre is
$$s = r\theta.$$

9. Find the lengths of the arcs of the following sectors of circles:
   (*a*) radius 3 cm, angle at centre 60°;
   (*b*) radius 5 cm, angle at centre 120°;
   (*c*) radius 8 cm, angle at centre 200°.

10. Assuming that the area of a circle is πr^2, show that the area A of a sector of angle θ radians is
$$A = \tfrac{1}{2}r^2\theta.$$

11. Calculate the areas of the sectors of Question 9.

12. An instrument pointer is 4 cm long. Its tip moves over a scale 5 cm long. Through what angle, in degrees, does it turn for a full-scale deflection?

241

3. ANGULAR VELOCITY

If a wheel rotates with a period of 5 seconds, then the angle θ which the radius through a fixed point P of the wheel makes with the horizontal is

$$\theta = \frac{2\pi}{5}\,t,$$

where t is the time in seconds measured from when P was at A.

The rate at which θ changes is

$$\frac{d\theta}{dt} = \frac{2\pi}{5}\text{ radians per second.}$$

This rate of change is called the *angular velocity* of the wheel.

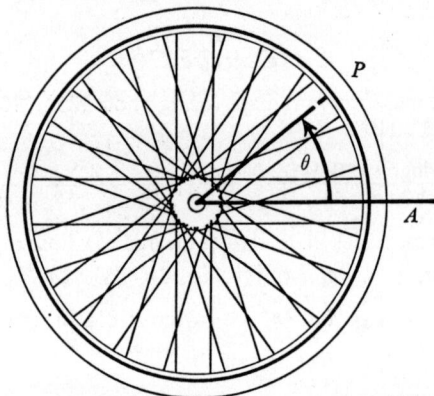

Fig. 10

Suppose the radius of the wheel is 1 m. Then the length of arc moved through by P is

$$s = 1\theta \text{ m} = \frac{2\pi t}{5} \text{ m.}$$

Hence, the velocity of P is

$$\frac{ds}{dt} = \frac{2\pi}{5} \text{ m/s.}$$

Example 1. Find the angular velocity of a 26 cm L.P. record making $33\frac{1}{3}$ revolutions per minute, and the velocity of a point on the rim.

The angle θ turned through in a time t seconds is

$$\theta = \frac{33\frac{1}{3} \times 2\pi}{60}\,t \text{ radians.}$$

Hence the angular velocity is

$$\frac{d\theta}{dt} = \frac{33\frac{1}{3} \times 2\pi}{60} \text{ radians per second}$$

$$= \frac{10\pi}{9} \text{ radians per second.}$$

Taking the radius as 13 cm, the velocity of a point on the rim is

$$13\frac{d\theta}{dt} = \frac{130\pi}{9} \text{ cm/s.}$$

Exercise D

1. Calculate the angular velocities of the second hand, the minute hand, and the hour hand of a clock.

2. Calculate approximately the velocity due to the earth's rotation of a man standing at:
    (*a*) the equator;   (*b*) latitude 60° N.
(Take the earth's radius as 6400 km.)

3. Find approximately the angular velocity of the moon about the earth. Taking its distance as 380000 km, calculate its velocity relative to the earth.

4. The angle θ turned through by a flywheel in t seconds is

$$\theta = t^2.$$

Find its angular velocity after 5 seconds.

5. An armature of an electric motor turns steadily at 2000 revolutions per minute. What is its angular velocity in radians per second?

4. THE GRADIENT FUNCTIONS OF SINE AND COSINE

4.1 The gradient of sin x. The result of Question 6, Exercise B, will have suggested that the gradient of the graph of the function $x \to \sin x°$ is proportional to $\cos x°$. If this is true it will follow that the graph of the function $x \to \sin x$, mapping x onto the sine of the angle whose radian measure is x, must also have a gradient proportional to $\cos x$. Further, since we deliberately introduced radian measure in order to make this gradient 1 when $x = 0$, and $\cos 0 = 1$, this gradient must everywhere be exactly equal to $\cos x$. This result we shall now formally prove. To do so we make use of the addition formulae (proved in Book T4, Chapter 6).

If $f: x \to \sin x$, then the gradient (derivative) of f at a is

$$f'(a) = \lim_{h \to 0} \frac{f(a+h) - f(a-h)}{(a+h) - (a-h)}$$

$$= \lim_{h \to 0} \frac{\sin(a+h) - \sin(a-h)}{2h}.$$

Now $\qquad\qquad\sin(a+h) = \sin a \cos h + \cos a \sin h,$

and $\qquad\qquad\sin(a-h) = \sin a \cos h - \cos a \sin h,$

so that $\qquad\sin(a+h) - \sin(a-h) = 2 \cos a \sin h.$

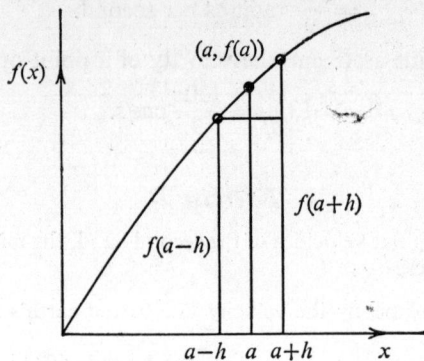

Fig. 11

Hence

$$f'(a) = \lim_{h \to 0} \frac{2 \cos a \sin h}{2h}$$

$$= \cos a \times \lim_{h \to 0} \frac{\sin h}{h}, \quad \text{since } \cos a \text{ does not vary with } h,$$

$$= \cos a, \quad \text{since the limit is 1.}$$

Hence, if $\qquad\qquad f(x) = \sin x, \quad f'(x) = \cos x.$

We can also obtain the derivative of $\cos x$ by this method (the details are left as an exercise), or directly from that for $\sin x$.

For $\qquad\qquad\qquad \cos x = \sin(x + \tfrac{1}{2}\pi);$

hence if $\quad f: x \to \cos x, \quad f'(x) = \cos(x + \tfrac{1}{2}\pi) = -\sin x.$

(We have here used the fact that translating by a constant parallel to the x-axis cannot alter the gradient.)

4.2 Linear approximations to $\sin x$ and $\cos x$. If x is small, we know that $(\sin x)/x \to 1$. It follows that x is the linear approximation for $\sin x$ when x is small. If the linear approximation for $\cos x$ is $1 + ax$, then we may use the identity $\sin^2 x + \cos^2 x = 1$ to find a. For

$$1 = \sin^2 x + \cos^2 x$$

$$= x^2 + (1+ax)^2 + \text{higher powers of } x$$

$$= x^2 + 1 + 2ax + x^2 a^2 + \text{higher powers}$$

$$= 1 + 2ax + \text{higher powers,}$$

so that $\qquad\qquad a = 0.$

244

This means that the linear approximation to cos x for small x is 1. We could equally well have obtained this by observing that the gradient of cos x is $-\sin x$, and that this is zero when $x = 0$.

4.3 Integrals of sine and cosine.

If $y = \sin x$, $\qquad\qquad \dfrac{dy}{dx} = \cos x.$

Hence $\qquad\qquad \int \cos x \, dx = \sin x + c.$

Again, if $y = \cos x$, $\qquad\qquad \dfrac{dy}{dx} = -\sin x.$

Hence $\qquad\qquad \int \sin x \, dx = -\cos x + c.$

4.4 Other circular functions.

Example 2. Find the gradient function of

$$x \cdot > \sin 2x.$$

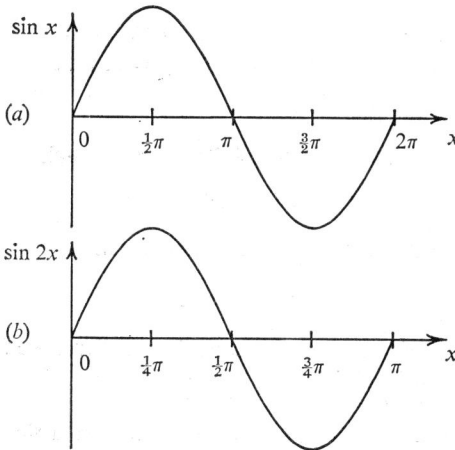

Fig. 12

Compare the graphs of sin x and sin $2x$ (Figure 12). Sin $2x$ is obtained from sin x, simply by rescaling the x-axis: $\frac{1}{2}\pi$ is replaced by $\frac{1}{4}\pi$, π by $\frac{1}{2}\pi$, and so on. Hence the gradient of all lines is *doubled*. Thus the gradient function of

$$x \to \sin 2x$$

is $\quad x \to 2 \cos 2x.$

Alternatively, we could work from first principles, as in Section 4.1.

Example 3. Find
$$\int \cos 2x \, dx.$$

Since
$$\frac{d(\sin 2x)}{dx} = 2 \cos 2x,$$

then
$$\int 2 \cos 2x \, dx = \sin 2x + c.$$

Hence
$$\int \cos 2x \, dx = \tfrac{1}{2} \sin 2x + c.$$

Exercise E

1. Find the gradient functions of:
   (a) $x \to \cos 2x$;   (b) $x \to \sin 3x$;   (c) $x \to \cos \tfrac{1}{2}x$;
   (d) $x \to \sin ax$;   (e) $x \to \cos ax$.

2. Integrate:
   (a) $\sin 2x$;   (b) $\cos 3x$;   (c) $\sin \tfrac{1}{2}x$;
   (d) $\cos ax$;   (e) $\sin ax$.

3. Find the gradient of dy/dx (written d^2y/dx^2) if
   (a) $y = \sin x$;   (b) $y = \cos x$;
   (c) $y = \sin ax$;   (d) $y = \cos ax$.

4. Show, from a graph, that
$$f: x \to \sin (x+c) \Rightarrow f': x \to \cos (x+c).$$
Deduce that $\int \cos (x+c) \, dx = \sin (x+c) + \text{constant}$.

5. Find the greatest and least values of $\sin x + \cos x$.

6. Find the area under one 'hump' of $\sin x$.

7. A particle moves on a straight line such that its distance s from a fixed point on the line at time t is given by
$$s = \sin t.$$
Find its velocity, and show that, if a is its acceleration at time t, then $a = -s$.
   Draw graphs to show the distance, velocity and acceleration, plotted against time.

8. The charge on a capacitor is given by
$$\int f(t) \, dt,$$
where the current i is given by $i = f(t)$.
   If $f(t) = \cos 100t$, and initially the charge is zero, calculate the charge after $\frac{1}{1000}$ s.

9. The back e.m.f. in an inductor is $50 \, di/dt$, where i is the current. What is its value after $\frac{1}{1000}$ s, if $i = \cos 100t$?

10. The gradient of a stretch of railway track is marked as 1 in 1320. What angle (in minutes) does it make with the horizontal?

11. If the sun subtends an angle of approximately 30′ at the earth, and its distance from us is $1 \cdot 5 \times 10^8$ km, what is its diameter?

12. If a pendulum is swinging through 10° on each side of the vertical, and takes 2 seconds to make a complete swing across and back, what is its angular velocity when it is passing through the vertical position? (Assume its angular displacement varies sinusoidally.)

① $\int_0^{\pi} \sin 2x \, dx$

② $\int_{-\pi/4}^{\pi/4} \cos x \, dx$

3 $\int_{\frac{\pi}{2}}^{0} (\sin x + \cos x) \, dx$

12

COMPUTATION AND LOGARITHMS

1. A NEW LOOK AT THE SLIDE-RULE

1.1 The slide-rule scale. By now we have become quite familiar with
the use of the slide-rule, and it is time to look more closely at the way in
which its scales are constructed. It will be sufficient to consider the C
(or D) scale to begin with (see Figure 1). The basic principle of the scale is
that the distance between any two marks depends only on the ratio of the
numbers marked there. Thus the distances from the mark 1 to the mark 2,
from 2 to 4, from 4 to 8, from 3 to 6, from 5 to 10, are all equal.

Fig. 1

Thus the distance $UB = 2 \times UA$, and $UC = 3 \times UA$. Denote the distance from U to the mark x by $d(x)$; then $UA = d(2)$,

$$UB = d(4) = d(2^2) = 2 \times d(2),$$

$$UC = d(8) = d(2^3) = 3 \times d(2)$$

and we could say for integral n that $d(2^n) = n \times d(2)$. Now 10 is not a whole-
number power of 2; it is more than 2^3 and less than 2^4, so that $d(10)$ is
between $3d(2)$ and $4d(2)$. If we note that $10^3 = 1000$ and $2^{10} = 1024$, we
see that $2^{10/3}$ is a little more than 10, so that $UT = d(10) = 3\cdot3d(2)$
approximately, and $d(2) = 0\cdot30d(10)$. It can be shown by a more precise
method that $d(2) = 0\cdot301d(10)$ to 3 significant figures, and accordingly
$d(4) = 0\cdot602d(10)$ and $d(8) = 0\cdot903d(10)$.

In terms of $d(10)$ as unit we can therefore construct the following table:

x	1	2	3	4	5	6	7	8	9	10
$d(x)$	0	0·301		0·602	0·699			0·903		1·000

We can insert $d(5)$ at once when we remember that

$$d(10) - d(5) = d(\tfrac{10}{5}) = d(2).$$

We can estimate $d(3)$ by noting that

$$d(81) = 4d(3) \quad \text{and} \quad d(80) - d(8) = d(10) \Rightarrow d(80) = 1\cdot903;$$

this gives $d(3) > 0\cdot476$ and in fact $d(3) = 0\cdot477$ to 3 significant figures. In this way the table can be built up and the numbers correctly located on the scale.

1.2   Logarithms. We call the number $d(x)$, in terms of $d(10)$ as unit, the *logarithm of x to base* 10, and we write it $\log_{10} x$. Because $d(10) = 1$, being our chosen unit, the values of $\log_{10} x$ for $1 < x < 10$ are all fractions between 0 and 1. It is clear that if the scale continued on the same principle we should have $d(10^2) = 2d(10)$, $d(10^3) = 3d(10)$, and in general

$$d(10^n) = nd(10) \text{ for integral } n;$$

or $$\log_{10} 10^n = n.$$

The logarithms of the numbers 1, 2, ..., 9 are therefore *indices* of a sort; they can be thought of as fractional powers of 10. Thus, approximately,

$$10^{0\cdot301} = 2, \quad 10^{0\cdot699} = 5,$$

and so on. The slide-rule principle enables us to *insert* these powers between the integral powers we know about already, that is, $10^1 = 10$, $10^2 = 100$ and so on. It should be clear that since $d(1) = 0$, being the distance from U to U, we must have

$$10^0 = 1 \quad \text{and} \quad \log_{10} 1 = 0.$$

1.3   The laws of logarithms are derived from the fundamental formula which defines what a logarithm is, that is,

$$x = 10^n \Leftrightarrow \log_{10} x = n.$$

Now if $10^n = x$ and $10^m = y$, then $xy = 10^n \times 10^m = 10^{n+m}$ when n and m are whole numbers, so that in this case

$$\log_{10} xy = n+m = \log_{10} x + \log_{10} y.$$

If we say that this equation is to hold when the logarithms are *not* whole numbers (and this is just what the slide-rule does) then we can also agree that $10^n \times 10^m = 10^{m+n}$, even when m and n are not whole numbers either.

This is usually called the first law of logarithms:

Law I.
$$\log x + \log y = \log xy.$$

If $y = 1$, then $\log_{10} x + \log_{10} 1 = \log_{10} x$, so that

$$\log_{10} 1 = 0, \quad \text{as before.}$$

If we put $xy = u$, then $y = u \div x$. Rearranging Law I, we have

$$\log(u \div x) = \log y = \log (xy) - \log x.$$

Law II. $$\log (u \div x) = \log u - \log x.$$

From Law I, we also deduce that, if p is a whole number,

$$\log (x^p) = \log (x \times x \times ...) \quad \text{with } p \ x\text{'s in the bracket}$$
$$= \log x + \log x + \log x + ... \text{ to } p \text{ terms}$$
$$= p \log x.$$

If we have defined $\log x$ by the 'slide-rule' process, we are at liberty to *define* x^p for all real p by this equation; that is,

'x^p is the number whose logarithm is $p \log x$'.

This is usually called the third law.

Law III. For all p, $\qquad \log x^p = p \log x.$

With this definition, 10^n will mean (for all n) that number whose logarithm is $n \log 10$. Since $\log_{10} 10 = 1$ (by definition), we have $\log_{10} 10^n = n$, which agrees with what we said at the outset.

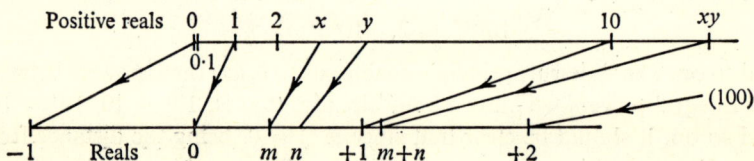

Fig. 2

It amounts to this: there is a continuous mapping from the positive real numbers onto the real numbers, such that if $x \to m$ and $y \to n$ then $xy \to m + n$ (see Figure 2).

In the mapping $x \to m$, $m = \log_{10} x$; in the inverse mapping $m \to x$, $x = 10^m$.

Once we have proved that the mapping exists, we may define it in either direction, and the inverse mapping can be deduced.

Exercise A

1. Copy and complete the table for $\log_{10} x$ as follows:

x	1	2	3	4	5	6	7	8	9	10
$\log_{10} x$	0	0·301		0·602	0·699			0·903		1·000

(a) $3^{13} = 1594323$ which is just under 1600000. Show that

$$\log_{10} 1600000 = \log_{10} 16 \times 10^5 = 6·204,$$

and hence estimate $\log_{10} 3$.

(b) From this, fill in the values of $\log_{10} 6$ and $\log_{10} 9$.

(c) Find $\log_{10} 48$ $(= \log_{10} 3 + 2 \log_{10} 4)$ and $\log_{10} 50$, and hence estimate $\log_{10} 49$ and $\log_{10} 7$.

2. Using the values in the table, compute the values of $\log_{10} 20$, $\log_{10} 24$, $\log_{10} 25$, $\log_{10} 28$.

3. Simplify:

(a) $\log (ab) - \log b$;   (b) $\log a^2 - \log a$;   (c) $\log_{10} (\frac{1}{4}) + \log_{10} 4$;

(d) $\log_{10} (\frac{2 \cdot 5}{4}) + \log_{10} 16$;   (e) $(\log x^3) \div (\log x)$;   (f) $\log_{10} 10^7$;

(g) $10^{\log_{10} 7}$;   (h) $3 \log_{10} 5 + 3 \log_{10} 2$.

4. Given $\log_{10} 2 = 0 \cdot 301$, what is $\log_{10} 2^{33}$? Suppose a single bacterium divides into two every hour, and each of its offspring also divides into two every hour, and so on; after 33 hours will there be more or less than ten thousand million bacteria present?

2. THE LOGARITHM TABLE

2.1   Use of the tables. The table gives logarithms to base 10 to three or four figures for numbers between 1 and 10. Numbers are multiplied by adding their logarithms (Law I), and then finding in the table the number whose logarithm is the sum.

Example 1. Multiply $3 \cdot 83 \times 2 \cdot 15$, using logarithms.

	No.	Log.
	$3 \cdot 83 \rightarrow$	$0 \cdot 5832$
We set the work out in two columns, thus:	$2 \cdot 15 \rightarrow$	$0 \cdot 3324$
	$8 \cdot 233$	$0 \cdot 9156$

We give our answer as $8 \cdot 23$, to 3 s.f.

Example 2. Divide $9 \cdot 45$ by $4 \cdot 08$.

	No.	Log.
Here we subtract the logarithms,	$9 \cdot 45$	$0 \cdot 9754$
and the answer is	$4 \cdot 08$	$0 \cdot 6107$
$2 \cdot 32$ to 3 s.f.	$2 \cdot 316$	$0 \cdot 3647$

Example 3. Evaluate $(1 \cdot 77)^4$ using logarithms.

	No.	Log.
Here we use Law III and multiply the	$1 \cdot 77$	$0 \cdot 2480 \times$
logarithm by 4. The answer is		4
$9 \cdot 82$ to 3 s.f.	$(1 \cdot 77)^4$	$0 \cdot 9920$

2.2   Numbers outside the range 1–10. If we write a number such as 257 in standard form, we have $257 = 2 \cdot 57 \times 10^2$. Use of the logarithm laws now shows that

$$\log_{10} 257 = \log_{10} \{2 \cdot 57 \times 10^2\}$$
$$= \log_{10} 2 \cdot 57 + \log_{10} 10^2 \quad \text{Law I}$$
$$= \log_{10} 2 \cdot 57 + 2 \log_{10} 10 \quad \text{Law II}$$
$$= 0 \cdot 4099 + 2$$
$$= 2 \cdot 4099.$$

We see that the log of 257 has the same decimal part as log 2·57 (sometimes this decimal part is called the *mantissa*—your slide-rule may have a scale on it marked MAN., if so, this is what it reads). The whole number part of log 257 (called the *characteristic*) is the number of places the figures are shifted to the left of the standard form position. Thus:

Number	Characteristic
2·57	0 (standard position)
25·7	1
257·0	2
2570·0	3

and so on, the decimal part in all cases being 0·4099.

If the figures are shifted to the *right*, the characteristic is a *negative integer*. Since it is convenient to keep the decimal part positive, we write the minus sign above the integer, thus: $\bar{3}$.

$$\log_{10} 0\cdot0257 = \log_{10} (2\cdot57 \times 10^{-2})$$

$$= 0\cdot4099 - 2$$

$$= \bar{2}\cdot4099, \text{ where only the 2 is negative.}$$

We can now continue our table the other way:

Number	Characteristic
2·57	0
0·257	$\bar{1}$
0·0257	$\bar{2}$
0·00257	$\bar{3}$

and so on. In all cases the characteristic is simply the number of places the figures have been moved from the standard position (between 1 and 10); + if the move is to the left, − to the right.

Manipulation of negative characteristics calls for a little care.

To add
$$\bar{1}\cdot3$$
$$\bar{2}\cdot8$$
$$\overline{\bar{2}\cdot1}$$

we say $3+8 = 11$; 1 (carried) $+\bar{1}+\bar{2} = \bar{2}$.

To subtract
$$\bar{1}\cdot3$$
$$\bar{2}\cdot8$$
$$\overline{0\cdot5}$$

we say $13-8 = 5$; and then *either* $1+\bar{2} = \bar{1}$, which subtracted from $\bar{1}$ leaves 0 (equal additions to top and bottom lines), *or else* $\bar{1}-1 = \bar{2}$, from which $\bar{2}$ is subtracted (i.e. +2 is added) to leave zero (decomposition of the top line).

252

Exercise B

1. Perform the following:

(a) 0·7+
    $\bar{2}$·5

(b) 0·7−
    $\bar{2}$·5

(c) 0·7−
    1·5

(d) $\bar{2}$·8+
    $\bar{1}$·7

(e) $\bar{2}$·8−
    $\bar{1}$·7

(f) $\bar{1}$·7−
    $\bar{2}$·8

(g) 2·3−
    3·1

(h) 2·3−
    3·9

(i) $\bar{2}$·3+
    $\bar{3}$·1

(j) $\bar{2}$·3+
    4·8

(k) $\bar{1}$·6+
    0·5

(l) 0·1−
    $\bar{1}$·9

2. Multiply $\bar{2}$·7 by 3, and write the answer in the same form.

To divide a 'mixed number' like $\bar{2}$·7 by 3, we have to remember that only every third integer ... $\bar{9}$, $\bar{6}$, $\bar{3}$, 0, 3, 6, 9 ... is divisible by 3, and we have to arrange always to have a positive remainder. Thus to work out 4·8 ÷ 3, we consider (3+1·8) ÷ 3 = 1·6, and to work out $\bar{2}$·8 ÷ 3 we consider ($\bar{3}$+1·8) ÷ 3 = $\bar{1}$·6. Figure 3 shows the positions of 4·8 and $\bar{2}$·8 on the number-line, the integers into which 3 will divide exactly, and the positive remainders. It should be clear that the integer to be divided is always the next divisible integer to the *left* (on the number-line) of the number to be divided.

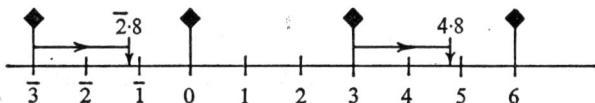

Fig. 3

3. Perform the following divisions:

(a) 2 | $\bar{1}$·8;

(b) 4 | $\bar{3}$·6;

(c) 3 | $\bar{1}$·7;

(d) 2 | $\bar{2}$·4;

(e) 2 | $\bar{4}$·5;

(f) 3 | $\bar{4}$·2;

(g) 5 | $\bar{1}$·5;

(h) 4 | $\bar{2}$·8;

(i) 5 | $\bar{2}$·7.

4. Evaluate by logarithms:

(a) 3·25 × 8·43;

(b) 131 × 247;

(c) 17·8 ÷ 27·2;

(d) 0·0237 × 12·4;

(e) 0·157 ÷ 0·0029;

(f) 78·04 ÷ 0·0341.

5. Evaluate the following powers and roots. (You multiply the log to get a power; what should you do for the corresponding root?)

(a) (32·7)^2;

(b) (1·98)^3;

(c) $\sqrt{(13·5)}$;

(d) (0·0184)^3;

(e) $\sqrt[3]{(0·25)}$;

(f) (1·17)^5.

6. Use logarithms to calculate:

(a) $\dfrac{2·65 \times 0·057}{198·3}$;

(b) $\sqrt{\left(\dfrac{2·93}{\pi}\right)}$;

(c) $\sqrt{\left(\dfrac{100 \times 3}{4\pi}\right)}$;

(d) (1·88)$^{1·4}$.

253

*2.3 The computation of tables. We constructed a rough table of logarithms by considering powers of 2, whose logarithms to base 2 are immediately obvious. Thus we have the simple table

Number	1	2	4	8	16	...	1024
$\log_2$	0	1	2	3	4	...	10

Now, just as in the case of the slide-rule the size of the scale does not really matter, so in the case of logarithms we may multiply them all by the same constant and they will work just the same. Then, since $2^{10} = 1024 = 10^3$ approximately, we can multiply all the logarithms by 0·3 and bring log 10^3 to the value 3, and log 10 to the value 1. By so doing we construct logarithms to the base 10. We are in fact using a particular case of a general rule, that logarithms to different bases only differ by a constant scale-factor. We can prove this formally as follows:

Suppose
$$x = a^m, \quad \text{then} \quad \log_a x = m.$$

Now $\log_b x$, where b is a different base, is given by
$$\log_b x = \log_b (a^m) = m \log_b a$$
by Law III.

Hence
$$\log_b x = \log_a x \times \log_b a,$$

and this last factor is a constant depending only on the two bases. This is sometimes called Law IV.

In our case above where the bases were 2 and 10, we have
$$\log_{10} x = \log_2 x \times \log_{10} 2 \quad \text{and} \quad \log_{10} 2 = 0.3$$

approximately, by our crude methods. More refined methods give
$$\log_{10} 2 = 0.30103.$$

Practical construction of logarithm tables depends on this same principle, but a much smaller factor than 2 is used. This gives numbers much nearer together than powers of 2. For example, we could take 1·01 as base. Then our table would be

Number	1	1·01	1·0201	1·0303	...	2·006	...	10·06
Logarithm	0	1	2	3	...	70	...	232

We have now only to divide all these logarithms by 232 to obtain the logarithms to base 10 of all these numbers—233 of them, quite close together. It will not now be difficult to fit the logarithms of two-figure numbers in between them.

Napier, who first discovered logarithms in 1614, used a method basically equivalent to this with a factor of 0·9999999. (Napier's logarithms at first went the opposite way to ours; they *increased* as the numbers *decreased*.

Briggs, who published the first tables in 1617, changed to our present system.) Another method proposed by Napier to obtain logarithms to ten places was to raise 2 to the power 10^{10} and count the digits! He stated that this number is $3\,010\,299\,957$, so that

$$0\cdot3010299956 < \log_{10} 2 < 0\cdot3010299957, \quad \text{which is correct;}$$

but it is not clear how he arrived at this answer.

The slide-rule was invented by Gunter in 1624. A modern slide-rule is constructed with the use of a table of logarithms. The kernel of the process is a long rod with a screw thread cut on it throughout its length to a high standard of precision. A master plate is mounted on a table bolted to a nut which is advanced by this screw (see Figure 4). The table moves under

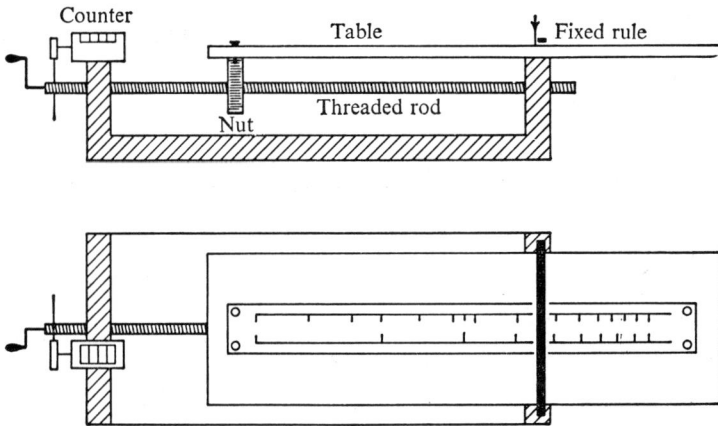

Fig. 4

a fixed rule as the crank is turned. A counter registers the advance of the screw. The process is simply that the master is advanced from the zero position (marked 1) by the crank; the logarithms of the numbers to be inscribed are read from the counter, and the scale divisions are scribed on the master plate by means of a scriber guided by the fixed rule. All depends on the accuracy of the screw thread and the gears which drive the counter. Modern slide-rules are made in plastic which is cut by dies made photographically by reduction from the master-plate.

3. PROGRAMMING

If you were asked to evaluate 2^7, you would probably say something like this:

'2,   4,   8,   16,   32,   64,   128'

and you would probably keep a count on your fingers of how many terms

255

of the sequence you had recited. You would then have two numbers stored—in your mind and on your fingers—at any moment, which we might call the 'reciter' and the 'counter'.

You can probably express the process in terms of a flow diagram, like this:

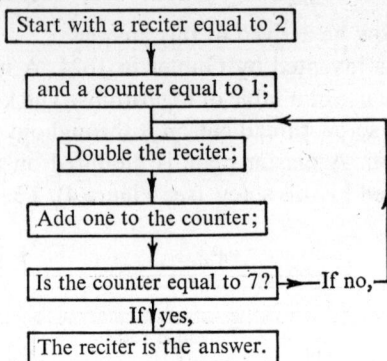

```
            ┌──────────────────────────────┐
            │ Start with a reciter equal to 2 │
            └──────────────────────────────┘
                          │
            ┌──────────────────────────────┐
            │ and a counter equal to 1;       │
            └──────────────────────────────┘
                          │               ◄──────────┐
            ┌──────────────────────────┐             │
            │ Double the reciter;        │             │
            └──────────────────────────┘             │
                          │                           │
            ┌──────────────────────────┐             │
            │ Add one to the counter;    │             │
            └──────────────────────────┘             │
                          │                           │
            ┌────────────────────────────┐──── If no,─┘
            │ Is the counter equal to 7?  │
            └────────────────────────────┘
                   If │ yes,
            ┌──────────────────────────────┐
            │ The reciter is the answer.     │
            └──────────────────────────────┘
```

You are using the reciter and the counter exactly as a computer uses its number stores, because when the reciter has reached (say) 32, it is of no further interest that you reached it by going through the sequence 2, 4, 8, 16. As soon as a new number is stored, the previous number held in that store may be forgotten.

Now, computers have to be programmed in special languages, designed so that the instructions fed in should be as economical as possible in symbols; and it is a programmer's job to reduce the particular problem to a set of instructions which the computer can obey. We shall tidy up the rather clumsy flow diagram above by using a convention which can easily be converted into any of the more usual languages in use today, and which, being based on such a language, is very much more economical, and also very much easier to understand.

We shall put our instructions in terms of the *addresses C* and *R* of the two stores, called 'the counter' and 'the reciter' in the diagram above, in three main ways.

(1) Instructions such as $\boxed{R:=2}$ are used to assign numbers directly to the stores. We read this as 'R is set equal to 2', or, more briefly, 'R becomes 2'.

(2) Instructions such as $\boxed{R:=R\times2}$ are used to assign numbers to the stores, this time using in the process numbers at present held in one of the stores. This instruction means, effectively,

$$\text{new reciter} = \text{previous reciter} \times 2,$$

and can therefore be used to replace the instruction 'double the reciter'.

256

(3) Decision boxes such as $\langle C = 7? \rangle$ are used to decide what action to take next. They should ask a 'yes-or-no' question about the sizes of the numbers in two stores, or, as here, compare the size of the number in one store with a particular number.

The flow diagram may then be written, more formally, like this:

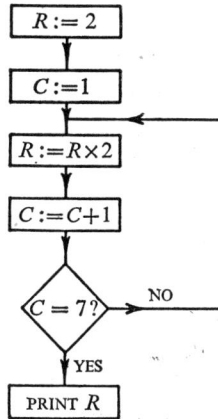

$$R := 2$$

$$C := 1$$

$$R := R \times 2$$

$$C := C + 1$$

$$C = 7?$$ NO

YES

$$\text{PRINT } R$$

Now, these instructions presuppose some sort of arithmetic unit, capable of taking pairs of numbers and combining them by the binary operations $+$, $-$, $\times$, and $\div$. But computers are built so that the numbers concerned can be specified simply by quoting their addresses; so that after the instruction

$$A := B + C$$

has been obeyed, the numbers stored in B and C are unchanged, while their sum is stored in A. Similarly, 'carbon copies' of the numbers in these stores are used for the simpler instructions $A := B$ and $A := -C$.

Thus, in a commercial program, customer A may buy (among other things) a certain number of bodgers. If the amount he owes the firm is stored in A, the price of a bodger in P, the number of bodgers in stock in S, and the number of bodgers he buys in N, the following part of a flow diagram will bring the firm's records up to date:

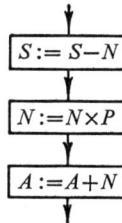

$$S := S - N$$

$$N := N \times P$$

$$A := A + N$$

so that stores A and S are now up to date, P still holds the price of a bodger for future reference, and N contains a useless number which will be superseded when the next purchase of bodgers is made.

Three other facilities are useful. First, besides the PRINT instruction (which displays the answer so that we can appreciate it), we often need a READ instruction. This takes numbers, in turn, from a list of data, and stores them in the stores named. Thus,

$$\boxed{\text{READ } A, B, C}$$

will take the first three numbers off the tape and store them in the stores whose addresses are A, B, and C respectively.

Secondly, powers occur so often that it is useful to have an instruction

$$\boxed{A := B\uparrow 7}$$

(rather than $A: = B^7$, which cannot be typed all on one line). The computer has a stored set of instructions like those in our original program, and this instruction brings them into play.

Thirdly, it is useful to be able to write more complicated instructions, such as

$$\boxed{A := (A\uparrow 3 - 3\times A + 1)\div 2}$$

Such an instruction will have to be reduced to a series of binary operations before a computer can obey it, but most computer languages are so designed that this reduction will be done for us in the computer itself. The brackets of course are used in the normal way to indicate when the usual conventions about the order of operations are over-ridden. These conventions are that $\uparrow$ is the operation taken first, then $\times$ and $\div$, and finally $+$ and $-$.

Exercise C

Write down the print-out resulting from the flow diagrams in Questions 1–4:

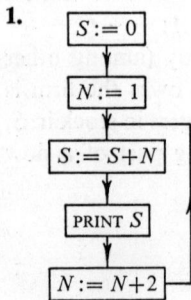

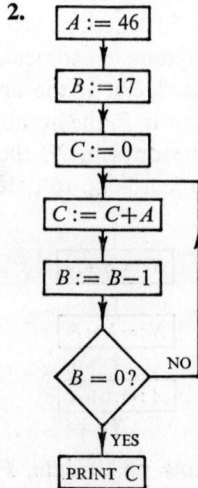

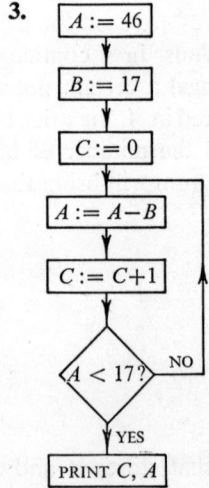

258

4.

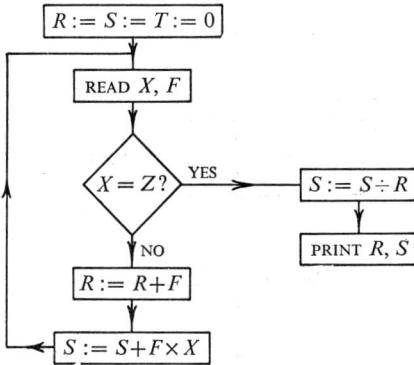

(In this program, the number stored in Z is printed, twice, at the end of the data list.)

Write flow diagrams to fulfil the following tasks:

5. Find the sum of the first 20 natural numbers.

6. Find $uv/(u+v)$, where u, v are two numbers on a data list.

7. Arrange three numbers on a data list in descending order.

8. Print out the largest number on the data list.

9. Evaluate $6 \times 1 \cdot 65^7$, without using the symbol ↑.

10. Write a given number greater than 0 in standard form (that is, in the form $r \times 10^m$, where m is an integer and $1 \leqslant r < 10$).

11. Print out a table of values, for $n = 1, 2, 3, \ldots$, of the sum of the cubes of the first n natural numbers.

12. Find the square root of 2 by trial and error, showing successively that $1^2 \leqslant 2 < 2^2, 1 \cdot 4^2 \leqslant 2 < 1 \cdot 5^2, 1 \cdot 41^2 \leqslant 2 < 1 \cdot 42^2$, etc., to four places of decimals.

4. ITERATION

4.1 Square roots.

Example 4. Suppose we want to find $\sqrt{10}$.

We could proceed like this: $3^2 = 9$, so 3 is too small. $\frac{10}{3}$ is therefore too large, but if we average these two we might be near the truth.

The average $= \frac{1}{2}(3+\frac{10}{3}) = \frac{19}{6}$.

$(\frac{19}{6})^2 = \frac{361}{36} = 10\frac{1}{36}$, so this is too large. $10 \div \frac{19}{6}$, that is $\frac{60}{19}$, is therefore too small; let us average again ... and so on.

We could set this out in a flow diagram.
To find $\sqrt{10}$

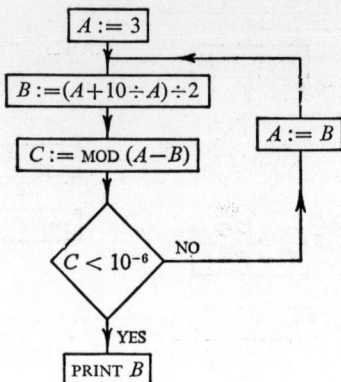

In this, MOD means 'take the modulus of'; i.e.

$$\text{MOD } (F) = \quad F \quad \text{if} \quad F \geqslant 0,$$
$$= -F \quad \text{if} \quad F < 0.$$

We go round the 'loop' until we get two numbers x and y, stored in A and B, which agree to, say, six places of decimals. This will be the required square root to this degree of accuracy.

Of course, how long this takes depends on our first guess. If we start with $x = 1$, we get, in order 1, 5·5, 3·659, 3·196, ... which is reaching the answer more slowly than if we start with $x = 3$, when we get successively 3, 3·167, 3·1623, 3·16228, ..., but we shall eventually arrive at any degree of accuracy we want.

This process of repeated 'trial and error', each result being used in the next trial, is called *iteration*; it is ideally suited to a computer, which is designed to carry out a simple arithmetical task over and over again. The process finds a sequence of values $x_1, x_2, x_3, \ldots$ approaching $\sqrt{10}$; the rule is that

$$x_{n+1} = \frac{1}{2}\left(x_n + \frac{10}{x_n}\right)$$

starting with $x_1 = 3$. We can replace 10 by any number N, and then the rule is

$$x_{n+1} = \frac{1}{2}\left(x_n + \frac{N}{x_n}\right).$$

4.2 Reciprocals. An amusing use of iteration is to find reciprocals without doing any division. If we have a computer with a fast multiplier it may be quicker to find a reciprocal this way than by long division. The

260

idea is this; suppose we want to find $\frac{1}{13}$. Then if x_n is a good guess, we want x_{n+1} to be a better one. $13x_n$ is nearly 1, and $13x_{n+1}$ must be nearer still. Which is larger, $1-13x_n$, or $1-13x_{n+1}$? Let us try putting the smaller equal to the square of the larger. (Why not the other way round?) Then $1-13x_{n+1} = (1-13x_n)^2$.

This gives
$$1-13x_{n+1} = 1-26x_n+169x_n^2,$$
$$x_{n+1} = 2x_n-13x_n^2,$$
$$= x_n(2-13x_n).$$

Let us start with
$$x_1 = 0 \cdot 1.$$

We obtain in succession
$$x_2 = 0 \cdot 07,$$
$$x_3 = 0 \cdot 07 \times 1 \cdot 09 = 0 \cdot 0763,$$
$$x_4 = 0 \cdot 0763 \times 1 \cdot 0081 = 0 \cdot 07691803,$$
$$x_5 = 0 \cdot 076918 \times 1 \cdot 000066 = 0 \cdot 076923077.$$

Actually $\frac{1}{13} = 0 \cdot 076923076923...$, so that we already have it correct to 9 decimal places.

Exercise D

1. Work out $\sqrt{5}$ by iteration, starting with $x_1 = 2$ and going as far as x_4.

2. Write out a flow diagram for finding $\frac{1}{13}$ to 6 decimal places by the method suggested.

3. Repeat Question 2 for $\frac{1}{7}$.

4. Find $\frac{1}{3}$ to 8 decimal places by using $x_{n+1} = x_n(2-3x_n)$, with $x_1 = 0 \cdot 3$.

5. Start with $x_1 = 10$ and find $\sqrt{99}$ to 4 SF. What does this give for $\sqrt{11}$?

6. Draw the graphs of (1) $x = 0 \cdot 2y$, (2) $y = 1 \cdot 7 - 0 \cdot 1x$. Show that the equations can be solved by the following procedure, and show what happens on the graph:

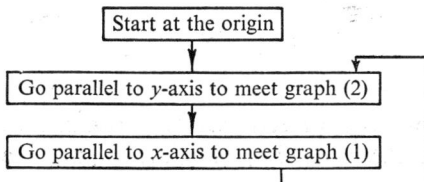

7. Show that the following flow diagram is equivalent to the procedure of Question 6, and use it to solve the two equations.

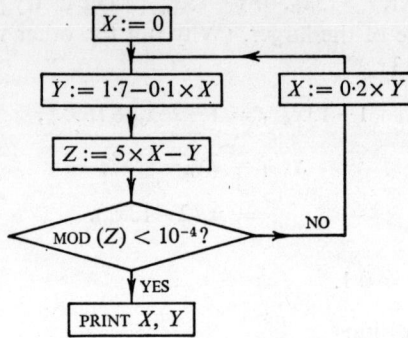

$$X := 0$$

$$Y := 1\cdot7 - 0\cdot1 \times X \qquad X := 0\cdot2 \times Y$$

$$Z := 5 \times X - Y$$

MOD $(Z) < 10^{-4}$? —— NO

↓ YES

PRINT X, Y

8. Show that if

$$x^3 = 2, \quad \text{then} \quad x = \frac{2}{3}\left(x + \frac{1}{x^2}\right).$$

Use the formula

$$x_{n+1} = \frac{2}{3}\left(x_n + \frac{1}{x_n^2}\right), \quad \text{with} \quad x_1 = 1$$

to obtain a sequence of three fractions approaching $\sqrt[3]{2}$.

9. What will be the output of the process in the following flow diagram? The first instruction reads a given number from the data tape.

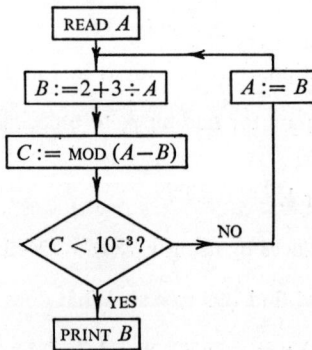

READ A

$$B := 2 + 3 \div A \qquad A := B$$

$$C := \text{MOD}\,(A - B)$$

$C < 10^{-3}$? —— NO

↓ YES

PRINT B

What happens when the given number is (*a*) 1, (*b*) 12, (*c*) -2, (*d*) 3, (*e*) -1?

5. NEWTON'S METHOD FOR SOLVING EQUATIONS

This is a very general type of iteration which can be used to find when any function of x takes the value 0, provided we can find the derived function. It is sometimes called the Newton–Raphson method.

Suppose we want to solve $f(x) = 0$. Imagine the graph of $f(x)$ drawn (see Figure 5), and that $f(\alpha) = 0$, so that α is the value of x we are looking for.

Suppose we make a 'near guess', $x = a$. $f(a)$ is represented by AP on the graph. If we draw the tangent at P and it meets the x-axis at T, then T will usually be nearer the point $x = \alpha$, where the graph crosses the axis, than A is. How do we find T?

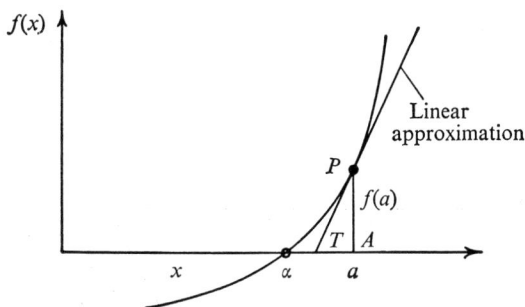

Fig. 5

PT is the linear approximation to the function at P. Now this we know from Chapter 9 to be

$$y = f(a) + (x-a)f'(a).$$

To find where this meets the x-axis we put $y = 0$ and obtain

$$f(a) + (x-a)f'(a) = 0$$

$$\Rightarrow x = a - \frac{f(a)}{f'(a)}.$$

This will be a nearer estimate than $x = a$, and of course, having found it, we may repeat the process to get a still nearer approximation.

Example 5. Find the positive root of

$$x^2 - x - 1 = 0.$$

Let $$f(x) = x^2 - x - 1.$$

Then $$f'(x) = 2x - 1.$$

Also $$f(1) = -1, \quad f(2) = 1,$$

so we expect that $f(x) = 0$ at some point between 1 and 2 (see Figure 6).

Start with $$x = 1, \quad f(1) = -1, \quad f'(1) = 1.$$

The next estimate is $$x = 1 - \frac{(-1)}{1} = 2.$$

Now $$f(2) = 1, \quad f'(2) = 3.$$

The next estimate is $$x = 2 - \tfrac{1}{3} = 1 \cdot 67.$$

Now $\qquad f(1\cdot67) = 0\cdot111,$

so that we next try $\qquad f'(1\cdot67) = 2\cdot33,$

$$x = 1\cdot67 - \frac{0\cdot111}{2\cdot33} = 1\cdot67 - 0\cdot05 = 1\cdot62,$$

and so on. [In fact $x = \frac{1}{2}(1 + \sqrt{5}) = 1\cdot618\ldots.$]

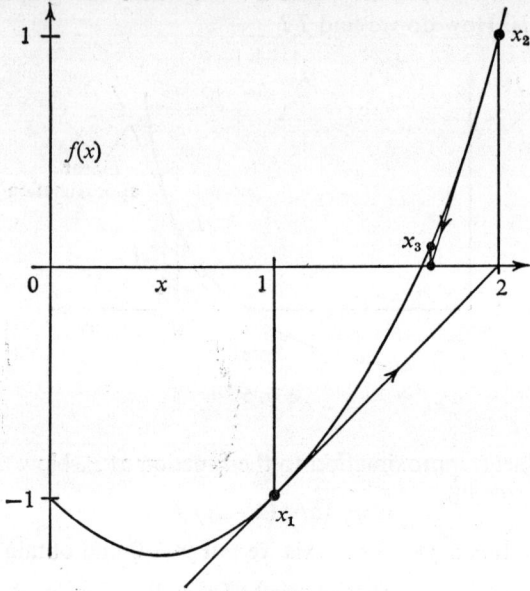

Fig. 6

We can write this process as that of computing a sequence $x_1, x_2, \ldots,$ where

$$x_{n+1} = x_n - \frac{f(x_n)}{f'(x_n)};$$

or as a flow diagram:

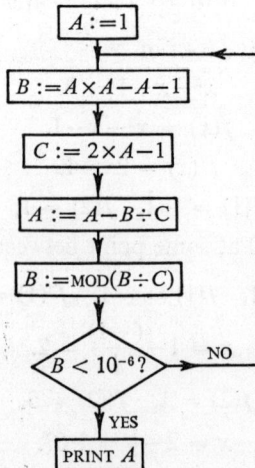

Exercise E

(Carry out these iterations to 2 decimal places.)

1. Find the root between 3 and 4 of $x^2 - 2x - 5 = 0$ by Newton's method. Start with $x = 3$.

2. Solve $x^2 - 5 = 0$ by Newton's method and compare with Question 1 of Exercise D. Is the sequence the same?

3. What happens if you use Newton's method to solve $2x - 5 = 0$ starting with $x = 2$?

4. Show that the sequence given by Newton's method for solving $x^3 - 2 = 0$ starting with $x = 1$, is the same as that in Question 8 of Exercise D.

5. Use Newton's method to solve $x^3 + x - 1 = 0$ to two places of decimals. Illustrate from the graph of $f(x) = x^3 + x - 1$.

6. Draw the graph of $y = \frac{1}{10}(15 - x^3)$ from $x = 0$ to 3.

Show that the x-coordinate of the point where it meets the line $y = x$ is a root of the equation $x^3 + 10x = 15$.

Find this root to 2 decimal places by Newton's method and check from your graph. Start with $x = 1$.

7. Find the solution of Question 6 by using the iteration formula

$$x_{n+1} = \tfrac{1}{10}(15 - x_n^3) \quad \text{and} \quad x_1 = 1.$$

[Figure 7 shows a graphical equivalent of this method. Discuss it.]

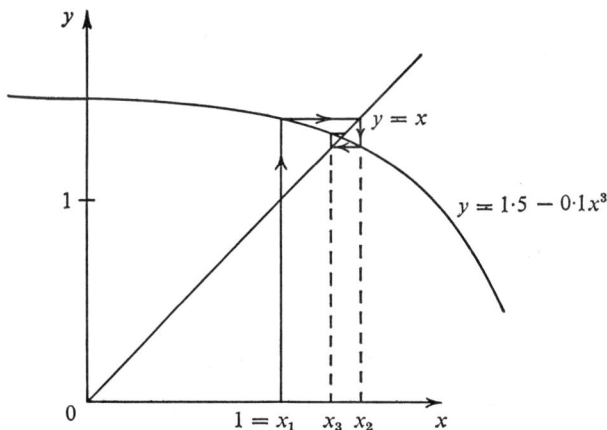

Fig. 7

8. Write a program for the solution of the equation

$$2 \sin x = x, \quad \text{where } x \text{ is measured in radians,}$$

to five decimal places. (Assume the machine can obey the orders $\boxed{A := \sin B}$ and $\boxed{A := \cos B}$.)

How many roots are there? Do you need a separate program for each? (Consider the graphs of $2 \sin x$ and of x.)

*9. A circular slice of bread is to be divided by two parallel cuts into three equal portions (see Figure 8). Show that if angle $POQ = x$, then $x - \sin x = \frac{2}{3}\pi$.

Write a program to solve this equation, and carry it through with the aid of tables to obtain x to the nearest degree.

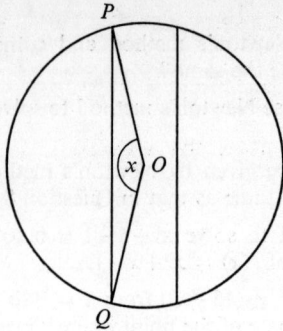

Fig. 8

*10. Show by a diagram that Newton's method will usually work if we write $x_{n+1} = x_n - [f(x_n)]/g$, where g is a fixed estimate of the gradient near the root. Obtain the solution of $x^3 - x - 5 = 0$ by this method, starting with $x_1 = 2$ and $g = 10$. Compare this with the standard method for speed.

13

FURTHER FUNCTIONS

1. COMPOSITE FUNCTIONS

1.1   A simple case. Consider the function

$$h: x \to (x^2-1)^3.$$

We can think of this in two ways:

(a) We can 'multiply it out' and obtain

$$h(x) = (x^2-1)^3$$
$$= x^6-3x^4+3x^2-1.$$

This is a polynomial function of a familiar type. Its derived function maps x onto $h'(x)$, where

$$h'(x) = 6x^5-12x^3+6x$$
$$= 6x\,(x^4-2x^2+1)$$
$$= 6x\,(x^2-1)^2.$$

The form of this result suggests that it could have been obtained more simply by using another approach.

(b) We regard the function as being effected in two steps:

$$f: x \to x^2-1 = u \quad \text{and} \quad g: u \to u^3.$$

Each step is simpler by itself than the actual mapping h. In a similar way a draughtsman might prepare, say, a pen-and-ink sketch of a building directly, or he might first take a photograph of the building and prepare his sketch from that. Each process probably involves 'enlargement' and distortion, that is a scale-factor depending on the function representing the process.

In our case the scale-factors are

$$f'(x) = 2x \quad \text{and} \quad g'(u) = 3u^2$$

and we see that $h'(x) = f'(x).g'(u)$.

Since $h(x)$ is the result of g mapping the image of f, we have

$$h(x) = g\{f(x)\},$$

and we write this as $h(x) = gf(x).$

The single mapping gf, which is h, is called the *composite* function of g on f.

Exercise A

In Questions 1–7, $f: x \to x^2$, $g: x \to x^{-1}$, $h: x \to 3x$, $s: x \to \sin x$.

1. What is the image of $x = 3$ under the functions fg, gf, gh, hg?

2. What is the image of $x = \frac{1}{3}\pi$ under the functions hs, sh?

3. What are the functions fg, gf, gh, hg, hs, sh?

4. Is composition of functions commutative?

5. What are the functions $(fg)h$ and $f(gh)$? Do you think composition of functions is generally associative?

6. What is the function gg?

7. If $y = \sin(3/x^2)$, express the function $x \to y$ in terms of f, g, h, and s.

8. If $f: x \to 2x+3 (= u)$, $g: u \to \sin u (= v)$, and $h: v \to v^2$, what is $hgf(x)$?

9. Take the function $x \to \sqrt{(1+\cos^2 x)}$ and express it as the composite of three functions, following the pattern of Question 8.

10. Resolve the following functions into their simple elements as in Questions 8 and 9:

(a) $x \to \sin^2 2x$; (b) $x \to \sin 2x^2$; (c) $x \to 2\sin^2 x$;

(d) $x \to \dfrac{1}{x^2-1}$; (e) $x \to \sin\left(\dfrac{1}{x^3}\right)$; (f) $x \to 2^{-x^2}$.

1.2 The derivative of composite functions.

We begin by considering two linear functions,

$$f: x \to u, \quad \text{where} \quad u = ax+b \quad \text{and} \quad g: u \to cu+d.$$

It is obvious in this case that $f'(x) = a$, $g'(u) = c$, and direct substitution gives

$$gf(x) = c(ax+b)+d = cax+bc+d.$$

Hence $\quad (gf)'(x) = ca = f'(x) \cdot g'(u).$

It is instructive to look at this graphically (see Figure 1). Since these are linear functions, the average gradient for each is simply the constant gradient. For

$$f: x \to u, \quad \frac{\delta u}{\delta x} = f'(x) \quad \text{(Figure 1}a);$$

for $\quad g: u \to y, \quad \dfrac{\delta y}{\delta u} = g'(u) \quad$ (Figure 1b);

and finally for $\quad gf: x \to y, \quad \dfrac{\delta y}{\delta x} = \dfrac{\delta y}{\delta u} \times \dfrac{\delta u}{\delta x} = f'(x) \cdot g'(u).$

When the functions are not linear, this relation is still true for the average

268

gradients over suitable steps, and can be proved to be true for the true gradients themselves. That is

$$\text{if} \quad f: x \to u, \quad g: u \to y,$$

then
$$\frac{dy}{dx} = \frac{dy}{du} \times \frac{du}{dx},$$

or
$$(gf)'(x) = f'(x) . g'f(x).$$

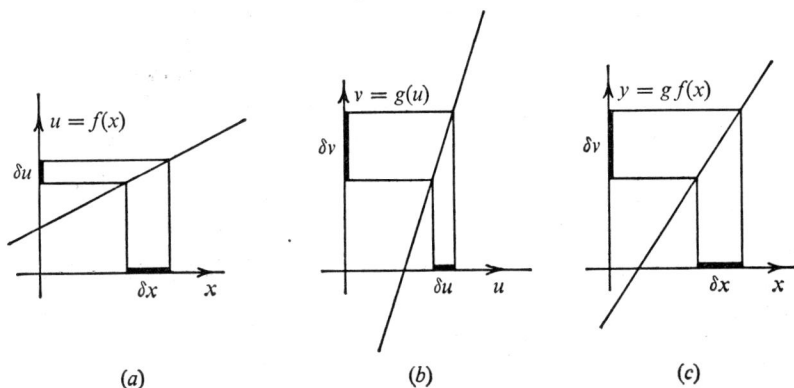

(a)   (b)   (c)

Fig. 1

Note that we cannot 'prove' that

$$\frac{dy}{dx} = \frac{dy}{du} \times \frac{du}{dx}$$

by 'cancelling the *du*' since *dy*, *du*, *dx* by themselves are meaningless. But the fact that the result is true makes the notation helpful. In this form it is often called the *Chain Rule*.

Example 1. Find $h'(x)$ if $h(x) = (x^2 - 1)^3$.

Here $h = gf$ where $f: x \to x^2 - 1$ and $g: x \to x^3$;

so that $f': x \to 2x$ and $g': x \to 3x^2$.

Hence $g'f: x \to 3(x^2 - 1)^2$,

so that $h'(x) = 6x(x^2 - 1)^2$.

Example 2. Find $h'(x)$ if $h(x) = \sin 5x$.

Here $h = gf$ where $f: x \to 5x$ and $g: x \to \sin x$;

so that $f': x \to 5$ and $g': x \to \cos x$.

Hence $g'f: x \to \cos 5x$,

so that $h'(x) = 5 \cos 5x$.

There is no need to go through the process of actually writing out both the functions every time one wants to differentiate a composite function. Example 1 is basically a 'cube' function and we differentiate it as such; but since it is the cube not of x but of x^2-1 we multiply the result by the derivative of x^2-1; i.e. $2x$.

Example 2 can be handled in the same way. It is basically a 'sine' function and we differentiate it as such; but since it is the sine not of x but of $5x$ we multiply by the derivative of $5x$, i.e. 5. A lot of time is saved by using this simple technique.

Exercise B

1. Find the derivatives of the following functions:

(a) $x \to \cos^2 x$; (b) $x \to \sin(2x-3)$; (c) $x \to (x-1)^5$;

(d) $x \to (x^3+x)^2$; (e) $x \to 3 \cos \pi x$; (f) $x \to \sin^2(\tfrac{3}{2}\pi x)$.

2. If $y = 3t^2$, $x = 2t$, find dy/dt, dx/dt. Express y as a function of x and find dy/dx. Does

$$\frac{dy}{dt} = \frac{dy}{dx} \times \frac{dx}{dt} \text{?}$$

3. If $f(x) = \sqrt[3]{x} = u$, and $g(u) = u^3$, what is $gf(x)$?
If $h = gf$, write down $h'(x)$, $g'(u)$. Use the formula $h'(x) = g'(u) \cdot f'(x)$ to find $f'(x)$.

4. Find $f'(x)$ when $f(x) = \sqrt{x}$ by a method similar to that of Question 3.

5. By giving values to t, plot the graph of the relation between y and x when $x = 20t$, $y = 30t-6t^2$, for $0 \leqslant t \leqslant 5$. Use the result verified in Question 2 to find dy/dx in terms of t.

6. The radius of a circular patch of mould is growing at the rate of 3 mm/day. How fast is the area growing when the radius is 10 mm?

$$\left[\text{Use } \frac{dA}{dt} = \frac{dA}{dr} \times \frac{dr}{dt}. \right]$$

*7. Sand is trickling through a hole and making a conical mound whose height is always equal to the radius of its base. The mound is 2 m high and its vertex is rising at 2 cm/min. How fast is the sand trickling onto it?

*8. The tide in a certain dock rises and falls in a simple sine wave with a period of 770 minutes. High tide is 4 m above low water mark. How fast is the water rising two hours after low water?

2. THE RECIPROCAL FUNCTION

The function $r: x \to 1/x$ is called the *reciprocal function*, and if

$$y = r(x) = 1/x,$$

y is said to be the reciprocal of x.

What is the reciprocal of y? What is $rr(x)$? The graph of $r(x)$ is shown in Figure 2. $r(x)$ is not defined when $x = 0$; the graph has a break there. When $x = 0\cdot01$, $r(x) = 100$; when $x = 0\cdot0001$, $r(x) = 10000$; the smaller x becomes, the larger $r(x)$ becomes. When x is negative, $r(x)$ is negative too; and again the nearer x is to zero on the negative side, the larger in numerical magnitude is the negative value of $r(x)$. Since

$$r(-x) = \frac{1}{-x} = \frac{-1}{x} = -r(x),$$

the graph is symmetrical about the origin;

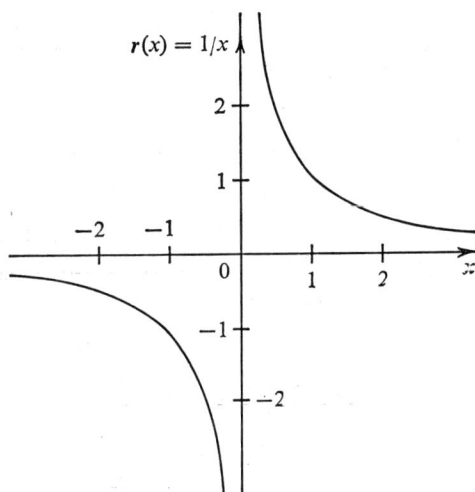

Fig. 2

since $\qquad\qquad y = \dfrac{1}{x} \Rightarrow x = \dfrac{1}{y},$

it is symmetrical about $x = y$ as well.

2.1   Derivative of $r(x)$. It is plain from the graph that $r'(x)$ is always negative. To find it, we consider the average gradient

$$\frac{r(x+h) - r(x)}{h} = \frac{1}{h}\left(\frac{1}{x+h} - \frac{1}{x}\right)$$

$$= \frac{1}{h} \cdot \frac{x - (x+h)}{x(x+h)} = \frac{-h}{hx(x+h)}$$

$$= \frac{-1}{x(x+h)}.$$

271

As h tends to zero, this tends to $-1/x^2$. Hence, if

$$r: x \to 1/x, \quad r'(x) = -1/x^2.$$

Since $x^2 > 0$ for all $x \neq 0$, this confirms that $r'(x) < 0$ for all $x \neq 0$, that is, for all values for which $r(x)$ is defined.

Example 3. An athlete runs 100 metres in 10 seconds. What can be said about his speed if the distance is accurate but the time may be $\frac{1}{10}$ s in error either way?

If the time is t seconds, the speed V is $100/t$ m/s. Hence if

$$f(t) = \frac{100}{t}, \quad \frac{f(t+h)-f(t)}{h^2} \simeq f'(t) = \frac{-100}{t^2}.$$

This gives

$$f(t+h)-f(t) \simeq -100h/t^2 = -0{\cdot}1 \quad \text{when} \quad h = 0{\cdot}1, \quad t = 10.$$

The speed is therefore between 9·9 and 10·1 m/s.

This is a particular case of a general and useful rule: the proportional small change in $r(x)$ is equal and opposite to the proportional small change in x.

To prove it, we have only to observe that

$$\frac{r(x+h)-r(x)}{r(x)} \simeq \frac{hr'(x)}{r(x)}$$

$$= \frac{-h/x^2}{1/x}$$

$$= -\frac{h}{x}.$$

In Example 3, a 1 % error in time makes a 1 % error in speed, but in the opposite sense.

Example 4. Differentiate

$$y = \frac{1}{x-2}.$$

$f: x \to \dfrac{1}{x-2}$ is a composite of two functions,

$$t: x \to u, \quad \text{where} \quad u = x-2 \quad \text{and} \quad r: u \to \frac{1}{u}.$$

Then $f(x) = rt(x)$ and

$$f'(x) = r't(x).t'(x) = -\frac{1}{[t(x)]^2} = -\frac{1}{(x-2)^2}.$$

The graph of $f(x)$ is merely that of $r(x)$ translated 2 units to the right (Figure 3).

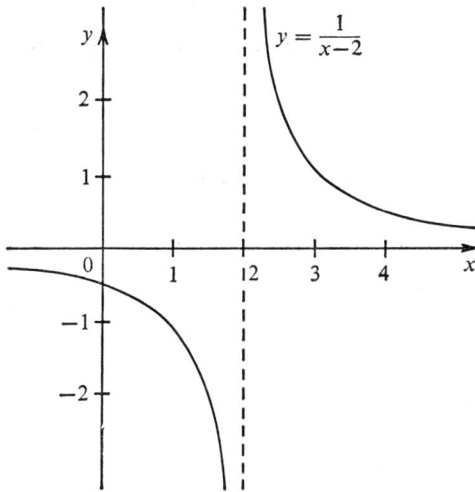

Fig. 3

Exercise C

1. Write down the derived functions of

(a) $x \rightarrow 5/x$;   (b) $x \rightarrow 1/(x+1)$;   (c) $x \rightarrow \dfrac{5}{x+1}$.

2. Sketch the graph of
$$y = \frac{5}{x+1}.$$

3. Find the gradient of $y = 12/x$ at $x = 4$ and the equation of the tangent to the curve there. Where does this line meet the two axes? Verify that its point of contact with the curve is midway between the points where it meets the axes.

4. Find the turning points of $y = x+(1/x)$. Which is a minimum point? Does either point give the least value of y?

5. Sketch the graphs of

(a) $y = x^2$;   (b) $y = -1/x$;   (c) $y = x^2-1/x$.

This last curve is called the *trident*. Where is its minimum point?

6. Boyle's law states that for a perfect gas in a container at constant temperature, the pressure p and the volume v are connected by an equation $pv = $ constant.

1000 cm³ of gas in a cylinder are under a pressure of 12 N/cm². If the volume is reduced by 30 cm³, the temperature remaining the same, find the rise in pressure.

7. A gang of twenty Young Consocialibs can pack up a batch of Election Manifestos in 5 hours. About how many minutes longer must they work if one fails to turn up?

273

3. INVERSE FUNCTIONS

We saw in Chapter 3 of Part 1 that if there is a relation xRy then y is said to be related to x by the *inverse relation*. If R is a function f, so that there is just one element y of the range which is related to a given element x of the domain, we write $f: x \to y$ and $y = f(x)$. When will the inverse relation also be a function?

Suppose f is the function $f: x \to x^2$. Then $f(3) = 9$ and $f(-3) = 9$ also. Accordingly, if R is the inverse relation, $9R3$ and $9R(-3)$; R is therefore not a function if negative numbers are included in the domain of f. On the other hand, for $f: x \to x^3$ with domain the real numbers, $f(3) = 27$ and there is no other x for which $f(x) = 27$. In this case, taking R as the symbol for the inverse relation, we could write $27R3$, and if $27Rx$ then $x = 3$. Since this is generally true, R is a function.

The distinction should be clear; if a function f maps just one element of its domain onto each element of its range, then there is an *inverse function* f^{-1} mapping the range of f back onto the domain of f. In this case f is said to be *one-one*.

For $f: x \to x^3$, $f^{-1}: x \to \sqrt[3]{x}$. But for $f: x \to x^2$, $f^{-1}: x \to \sqrt{x}$, which is only a function if we restrict both domain and range to positive numbers or zero.

3.1 Derivatives of inverse functions. Suppose we are in a moving car and have travelled x m in the t seconds since we started. There is a function $f: t \to x$. At the instant considered, our speed is

$$dx/dt = f'(t), = 6 \text{ m/s} \quad \text{(say)}.$$

But we could equally well have considered the situation from the point of view of distance covered and said that it had taken us t seconds to cover x m. There is a function $g: x \to t$, and its derivative $g'(x) = dt/dx$ is the rate at which time is changing with respect to distance. This is obviously $\frac{1}{6}$ second per metre; so that

$$\frac{dx}{dt} = 6\text{m/s}, \quad \frac{dt}{dx} = \frac{1}{6}\text{ s/m}.$$

It appears that $$\frac{dx}{dt} \times \frac{dt}{dx} = 1,$$

and this is generally true, as we shall show.

We use the method of Exercise B, Questions 3 and 4. If g is the inverse function of f, then $$y = f(x) \implies x = g(y).$$

Consider the composite function gf. Then $gf(x) = g(y) = x$, and the derived function $(gf)'$ must be the constant 1. Hence

$$1 = (gf)'(x) = g'(y).f'(x),$$

274

so that
$$g'(y) = \frac{1}{f'(x)}.$$

If we write
$$g'(y) = \frac{dx}{dy}, \quad f'(x) = \frac{dy}{dx},$$

then the result is equivalent to
$$\frac{dy}{dx} \times \frac{dx}{dy} = 1.$$

Again the notation suggests (though of course it does not prove) the truth of the result.

Exercise D

1. What are the inverses of the following functions?

(a) $x \to 3x - 2$;   (b) $x \to x^3 + 1$;   (c) $x \to 12/x$;

(d) $x \to 5 - x$;   (e) $x \to \dfrac{2x - 3}{x + 1}$;   (f) $x \to \dfrac{2x - 3}{x - 2}$.

2. If $pv = 1000$, verify that
$$\frac{dp}{dv} = \frac{1}{dv/dp}.$$

3. If $(x^m)^p = x$, and the index laws hold, then $mp = 1$. It is therefore convenient to denote the inverse function of $f: x \to x^m$ by $f^{-1}: x \to x^{1/m}$. Find the derivative of $f^{-1}(x)$. Does it obey the rule
$$f(x) = x^n \Rightarrow f'(x) = nx^{n-1}?$$

4. EXPONENTIAL FUNCTIONS

4.1 Derivatives of exponential functions. You may remember the old catch-question: 'A water-lily leaf doubles in size every week. After 10 weeks it fills the circular pond in which it is growing. How long did it take to half-fill the pond?' We are interested in a harder question, but there is no catch in it. How fast is its area growing at any time?

Suppose its area was 1 unit to begin with. After a week it would be two units, after 2 weeks, 4 units, and after t weeks, 2^t units. What is the derivative of $f: t \to 2^t$? Intuitively we should guess that a leaf twice as big has to grow twice as fast, and so on; that is, we suspect that $f'(t)$ is proportional to $f(t)$. Let us see if our conjecture is true.

The average gradient of f from t to $t + h$ is (Figure 4)
$$\frac{RM}{QM} = \frac{f(t+h) - f(t)}{h} = \frac{2^{t+h} - 2^t}{h}$$
$$= 2^t \times \left[\frac{2^h - 1}{h}\right].$$

Now this is a product of two terms; the first, 2^t, depends only on t, while

275

the second does not depend on t but only on h. Further, from Figure 4 we can see that this second term is PL/AL which is the gradient of the chord AP. Hence, as h tends to 0, this second term tends to the gradient at A, a constant which we will call K_2.

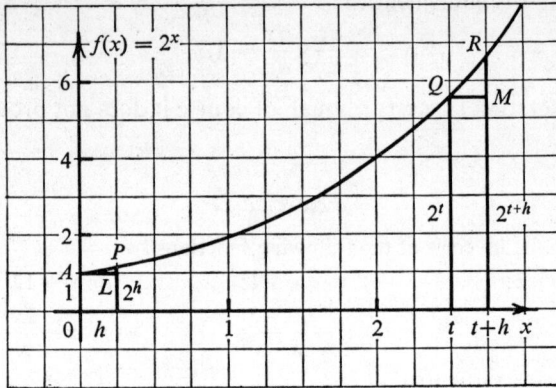

Fig. 4

Thus $f'(t) = K_2.2^t$, and our conjecture is confirmed. In the same way the derivative of 4^t will be $K_4.4^t$, where K_4 is the gradient of 4^t at $t = 0$, and so on. Since $4^t = 2^{2t}$, we can use the rule for composite functions and obtain the equivalent forms for the gradient

$$K_4.4^t = K_2.2^{2t} \times 2$$
$$= 2K_2.4^t,$$

so that $K_4 = 2K_2$, $K_8 = 3K_2$ and so on, the K's being related like the distances on a slide-rule. We saw in the previous chapter that this means that $K_n = \log n$ to some base.

By drawing the graphs of 2^x and 3^x we can estimate K_2 and K_3; we find (see Exercise E, Question 1) $K_2 \simeq 0.7$ and $K_3 \simeq 1.1$. There must therefore be a number e between 2 and 3 for which $K_2 = 1$; for this number

$$f(x) = e^x \Rightarrow f'(x) = e^x.$$

Further $K_n = \log_e n$, (since $\log_e e = 1$), and

$$f(x) = n^x \Rightarrow f'(x) = \log_e n.n^x.$$

It can be shown that e is an irrational number, whose approximate value is 2.718.

4.2 Some examples of exponential functions. In a given population, it would be reasonable to expect that the number of births and deaths in a fixed interval of time would be proportional to the size of the population.

276

Suppose that the population increases by a tenth each year. Thus, if the population at the beginning of a year was N, at the end of the year it would be approximately $11N/10$ and at the end of the next year approximately $(\frac{11}{10})^2 N$ and so on. (Why must we say 'approximately'?)

After T years the population would be $N(11/10)^T$; this formula cannot hold exactly, but as a first approximation we might consider that it is true even for non-integral values of T.

A similar formula can be derived from radioactive decay. Suppose that there are N radioactive atoms and that $\frac{1}{10}$ of these change into non-radioactive atoms in an hour. Then after t hours there will be $N(\frac{9}{10})^t$ atoms which are still in a radioactive state.

Example 5. A besieged garrison decides to eat $\frac{1}{12}$ of its remaining rations each day. How much remains after n days?

We argue as above: obviously $\frac{11}{12}$ of the original rations remains after the first day and $(\frac{11}{12})^n$ after n days.

We now look at a slightly harder example which uses the idea of *linear approximation* (see Chapter 9).

Example 6. A burglar estimates that his chance of being arrested in any particular week is $\frac{1}{15}$. What is his chance of being arrested in x weeks? Use linear approximation to estimate his chance of being arrested on a particular day.

[The gradient function for $(\frac{14}{15})^x$ is $-0{\cdot}069.(\frac{14}{15})^x.$]

His chance of not being arrested in any particular week is $\frac{14}{15}$, and thus his chance of not being arrested in x weeks is $(\frac{14}{15})^x$; and therefore his chance of being arrested is $1-(\frac{14}{15})^x$.

The linear approximation for $f(x)$ when $x = 0$ is given by

$$f(x) \simeq f(0) + xf'(0).$$

We now require $f(\frac{1}{7})$.

$$f(\tfrac{1}{7}) \simeq 0 - \tfrac{1}{7}.(-0{\cdot}069) \simeq 0{\cdot}01.$$

4.3   The logarithmic function. We saw in Chapter 12 that the inverse function to $f: x \to 10^x$ was $f^{-1}: y \to \log_{10} y$ $(y > 0)$. In the same way we may take e as the base of our logarithms and define the inverse function to $f: x \to e^x$ as $f^{-1}: y \to \log_e y$. This is a hopelessly inconvenient base for practical use, but of course Law III of logarithms tells us that $\log_e y$ and $\log_{10} y$ are simply proportional. The great advantage of e as base is seen when we consider the derived function. By the rule for inverse functions

$$y = e^x \Rightarrow x = \log_e y \quad (y > 0),$$

$$\frac{dy}{dx} = e^x \Rightarrow \frac{dx}{dy} = 1 \bigg/ \frac{dy}{dx} = 1/e^x = \frac{1}{y}.$$

277

The derived function of $\log_e$ is r, the reciprocal function, the domain being positive real numbers. This is a crucial result. The most important consequence is that it completes our integration table; for if

$$g(x) = \log_e x \Rightarrow g'(x) = \frac{1}{x},$$

then
$$\int^t \frac{1}{x} dx = g(t) = \log_e t + c \quad (t > 0).$$

Now $\log_e 1 = 0$ (since $e^0 = 1$, as in Chapter 12). We can therefore only have $c = 0$ if we start to integrate from $x = 1$. Indeed, to begin at $x = 0$ is impossible, since the area under the graph of $r(x)$ increases without limit as x tends to zero. But for any positive number t,

$$\int_1^t \frac{1}{x} dx = \log_e t$$

(see Figure 5).

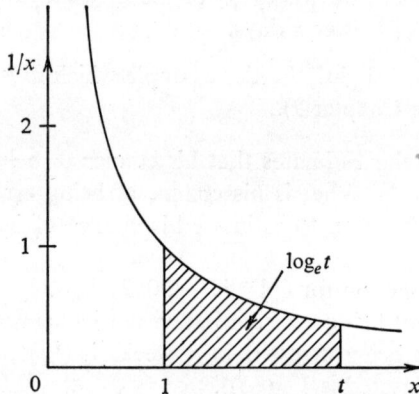

Fig. 5

Example 7. *Half-life.* A radioactive element emits particles of its own accord and thereby changes into some other element. This process is called *decay* and the rate at which it takes place is always proportional to the amount of element present. The time taken for half a given mass of radioactive substance to decay is called its *half-life*.

The half-life of radium is about 1600 years. Formulate a mathematical equation for its rate of decay and find (in mgm/year) the rate of decay of 1 gm of radium.

If m gm is the mass after t years, then $m = f(t)$ where

$$f(1600) = \tfrac{1}{2} f(0),$$

$$f(1600n) = (\tfrac{1}{2})^n f(0)$$

and thus
$$f(t) = (\tfrac{1}{2})^{t/1600} f(0).$$

$f(0)$ is the mass to begin with, say m_0.

278

Thus $\qquad\qquad\qquad m = m_0 \cdot k^t \quad$ where $k = (\tfrac{1}{2})^{1/1600}$.

Hence $\qquad\qquad\qquad \dfrac{dm}{dt} = m_0 \log_e k \cdot k^t$

$$= m \cdot \log_e k$$

$$= -\frac{\log_e 2}{1600} \cdot m.$$

Alternatively,

$$\frac{dm}{dt} = -pm, \quad \text{for some constant } p.$$

$$\Rightarrow \frac{dt}{dm} = -\frac{1}{p} \cdot \frac{1}{m}$$

$$\Rightarrow t = -\frac{1}{p} \log_e m + c, \quad \text{integrating,}$$

$$\Rightarrow t + 1600 = -\frac{1}{p} \log_e \left(\frac{m}{2}\right) + c, \quad \text{since the mass is halved}$$

$$\text{after 1600 years,}$$

$$\Rightarrow 1600 = -\frac{1}{p} \log_e (\tfrac{1}{2}), \quad \text{subtracting;}$$

$$= \frac{1}{p} \log_e 2.$$

Hence $\qquad\qquad\qquad p = \dfrac{\log_e 2}{1600}, \quad$ as before.

When

$$m = 1 \text{ gm}, \quad \frac{dm}{dt} = -p \text{ gm/yr} = -0.43 \text{ mgm/yr approximately.}$$

Exercise E

1. Draw the graphs of 2^x and 3^x for $-1 \leqslant x \leqslant 3$, and estimate their gradients when $x = 0$.

2. Differentiate

(a) $\log 3x$; $\qquad$ (b) e^{3x}; $\qquad$ (c) e^{-x^2}; $\qquad$ (d) 3^x.

3. What are the derivatives of $\log ax$ and $\log x$? Explain your result.

4. A function f has the property that $f(kx) = f(x) + f(k)$ for all positive numbers k. Deduce that

$$f'(kx) = \frac{1}{k} f'(x) \quad \text{and hence that} \quad f'(k) = \frac{1}{k} f'(1).$$

5. The proportion of the radioactive Carbon 14 in living wood does not vary with time, but for dead wood it decays exponentially with a half-life of 6000 years. Given a piece of wood whose radioactivity is 75 % of a similar living piece, how long ago did it die?

279

6. A population of flies increases by 50 % every week. Use linear approximation to estimate the percentage increase in an hour.

7. In a certain war a soldier's chance of being put out of action during 12 hours at the front is $\frac{1}{50}$. Find his chance of surviving for $12t$ hours. Find the chance that there is just one casualty in a regiment of 500 in $12t$ hours, and hence the most likely length of time for there to be just one casualty.

8. The chance of throwing six with a weighted die is $\frac{1}{5}$. Find the chance of throwing just one six in N throws and hence estimate the most likely number of throws in which you will throw just one six.

9. *Newton's law of cooling.* A body at temperature 100 °C stands in a room whose temperature is 10 °C. The temperature falls at a rate proportional to its excess above the temperature of the room. Take θ degC as this temperature excess and write this law in mathematical symbols. By integrating $dt/d\theta$ find how long it takes to cool to 20 °C if it begins to cool at a rate of 4·5 degC/min.

5. MISCELLANEOUS APPLICATIONS

5.1 Rates of change.

Example 8. The pressure p N/m² and the volume v m³ of a gas at constant temperature are connected by the relation

$$pv = 0·05.$$

If the volume is increasing at a rate of 5×10^{-6} m³/s, find the rate of change of pressure with time when $p = 1000$.

There is a function f mapping time onto volume: $v = f(t)$.

There is a function g mapping volume onto pressure

$$p = g(v) \quad \text{where} \quad g: v \to 0·05/v.$$

Hence there is a composite function mapping time onto pressure. The chain rule tells us that the rate of change of pressure with respect to time is equal to the rate of change of pressure with respect to volume multiplied by the rate of change of volume with respect to time. Thus

$$\frac{dp}{dt} = \frac{dp}{dv} \times \frac{dv}{dt}.$$

But $\qquad dp/dv = -0·05/v^2 \quad \text{and} \quad dv/dt = 5 \times 10^{-6}.$

Hence when $p = 1000$, $v = 5 \times 10^{-5}$, $dp/dv = -2 \times 10^7$, and

$$dp/dt = -100.$$

So that the rate of change of pressure is -100 N/m² s.

(The negative sign indicates that the pressure is decreasing.)

280

Example 9. (*a*) When a metal sphere is heated the radius expands by a small percentage α. What is the approximate change in volume?

If the volume is V, and the radius r, then

$$V = \tfrac{4}{3}\pi r^3$$

and

$$\frac{dV}{dr} = 4\pi r^2.$$

(Why is this equal to the surface area?)

If the radius increases by $\alpha \%$ its actual increase is

$$\frac{\alpha r}{100}.$$

Thus the approximate change in volume is given by $\dfrac{dV}{dr} \times$ the change in r

$$= 4\pi r^2 \times \frac{\alpha r}{100}.$$

Hence the approximate percentage change in volume is

$$\frac{\text{increase in volume}}{\text{original volume}} \times 100 = \frac{4\pi \alpha r^3}{\tfrac{4}{3}\pi r^3} = 3\alpha.$$

(*b*) Suppose on the other hand that the volume changes by a small percentage ϕ; what is the approximate change in the radius?

From Section 3

$$\frac{dr}{dv} = \frac{1}{dv/dr} = \frac{1}{4\pi r^2} = \frac{r}{3V}.$$

Hence the approximate change in r is given by $\dfrac{dr}{dV} \times$ the change in V

$$= \frac{r}{3V} \times \frac{\phi V}{100} = \frac{\phi r}{300}.$$

Thus, as V is increased by $\phi \%$, r will be increased by approximately $\tfrac{1}{3}\phi \%$, which is what we would expect from part (*a*).

5.2   Graphs and areas.

Example 10. Sketch the graph of the function $f: x \to 1/(x-1)^2$, and find the area between the *x*-axis, and the lines $x = 2$ and $x = 3$. The graph is obtained from that of $1/x^2$ by a translation of one unit to the right (Figure 6). For the area we first require a function A such that

$$A'(x) = 1/(x-1)^2.$$

The derivative of $1/x$ is $-1/x^2$, and that of $1/(x-1)$ is $-1/(x-1)^2$ and so

$$A(x) = -1/(x-1)+c.$$

Now for our particular area function we require that

$$A(2) = 0$$

so that $$c = 1$$

and $$A(3) = -\tfrac{1}{2}+1$$

$$= \tfrac{1}{2} \quad \text{which is the required area.}$$

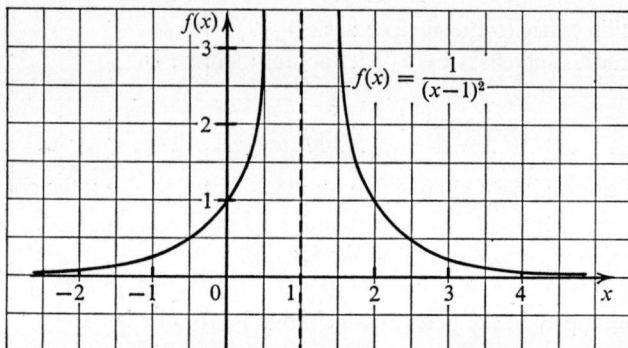

Fig. 6

Example 11. Sketch the graph of $f(x) = x/(x^2+1)^2$ and find the area between the curve and the x-axis from $x = 0$ to $x = 2$.

The graph has point symmetry about the origin. (This can be seen from the fact that $f(2) = -f(-2)$, etc.) $f(x)$ tends to 0 for large values of x and has no zeros other than $x = 0$. $f(x) > 0$ for $x > 0$, and $f(0) = 0$; there must therefore be a turning point for some positive x. A table of values gives a few points.

x	0	$\tfrac{1}{4}$	$\tfrac{1}{2}$	$\tfrac{3}{4}$	1	$1\tfrac{1}{2}$	2
$f(x)$	0	0·22	0·32	0·31	0·25	0·14	0·08

We note that $f(2)$ is very small and that $f(\tfrac{3}{4})$ and $f(\tfrac{1}{2})$ are roughly equal. This suggests that the turning point is about half-way between $x = \tfrac{1}{2}$ and $x = \tfrac{3}{4}$. The graph can now be sketched as in Figure 7.

To find the area we proceed as in Example 9. The derivative of $1/x$ is $-1/x^2$; the derivative of $1/(x^2+1)$ is $-2x/(x^2+1)^2$; and so

$$A(x) = -1/2(x^2+1)+c.$$

In this case we require that $A(0) = 0$, hence

$$c = \tfrac{1}{2}$$

and $$A(2) = -0·1+0·5$$

$$= 0·4.$$

It is important to note that we could not have integrated $1/(x^2+1)^2$ at this stage; but we could have integrated $x/(x^2+1)$ which is similar in form to $1/x$. At the moment we are able only to integrate a few special functions —simple powers of x, simple trigonometric functions and functions obtained by differentiating composite functions.

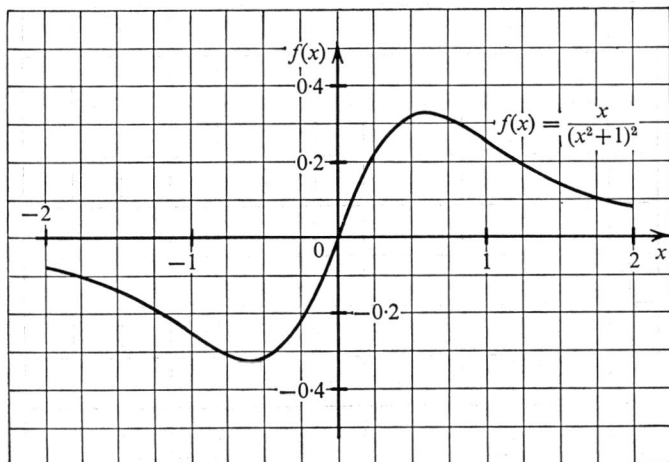

Fig. 7

5.3 Maxima and minima.

Example 12. What is the smallest amount of metal needed to make a cylindrical tin of volume 16π cubic centimetres?

Suppose the radius is x cm and the height l cm. Then the volume is $\pi x^2 l$ cm^3. Hence

$$\pi x^2 l = 16\pi$$

$$\Rightarrow l = \frac{16}{x^2}.$$

The area is $2\pi xl + 2\pi x^2$ cm^2 (neglecting the rims and the seam). Substituting for l, the area is A cm^2, where

$$A = 2\pi x \cdot \frac{16}{x^2} + 2\pi x^2$$

$$= \frac{32\pi}{x} + 2\pi x^2.$$

Figure 8 shows the graph of A (since the radius must be positive, it is sufficient to consider positive values of x only).

For the smallest value of the area the gradient of A has to be zero. The gradient function is

$$\frac{dA}{dx} = -\frac{32\pi}{x^2} + 4\pi x.$$

The gradient is zero when

$$-\frac{32\pi}{x^2}+4\pi x = 0, \quad \text{that is, when} \quad x = 2.$$

Then $l = 16/2^2 = 4$, and the smallest area is $16\pi+8\pi = 24\pi$ cm².

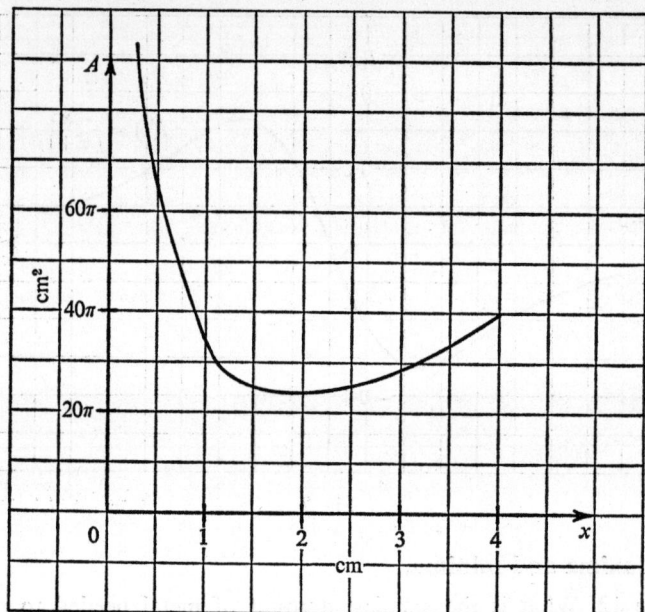

Fig. 8

Needless to say, this example was 'contrived' to make the answer simple. In a practical case, not only would the figures be awkward, but allowances for seams and wastage would radically alter the situation.

Miscellaneous Exercise

1. Consider the two functions with positive domains

$$f: x \to t, \quad \text{where} \quad t = x^5; \quad \text{and} \quad g: y \to t, \quad \text{where} \quad t = y^2.$$

Show that

$$y = g^{-1}(t) = g^{-1}f(x)$$

$$= \sqrt{x^5}, \quad \text{which is written } x^{\frac{5}{2}}.$$

Use the chain rule

$$\frac{dt}{dx} = \frac{dt}{dy}\times\frac{dy}{dx}$$

to find dy/dx in terms of x, and verify that it satisfies the rule

$$f(x) = x^n \Rightarrow f'(x) = n \cdot x^{n-1}.$$

284

2. Find the derivatives of:

(a) $x^{\frac{3}{2}}$; (b) $x^{\frac{5}{2}}$ $(x > 0)$.

What do you think $x^{-\frac{1}{2}}$ means, and what will its derivative be?

3. Find:

(a) $\int x^{-2}\,dx$; (b) $\int x^{-3}\,dx$; (c) $\int x^{\frac{3}{2}}\,dx$.

4. Find the gradient functions of the following:

(a) $(4x+1)^3$; (b) $(4x^2+1)^3$; (c) $(4x^3+1)^3$;

(d) $(x^2+3x+1)^2$; (e) $(x^2+3x+1)^3$; (f) $2/x^3$;

(g) $3/x^7$; (h) $1/(x+1)$; (i) $1/(x+1)^2$;

(j) $1/(2x^2+1)$; (k) $\cos 8x$; (l) $\sin^5 3x$.

5. Sketch the graphs of the functions in 4(i) and 4(j).

6. The acceleration of a particle moving in a straight line is given by $a = 1/t^4$. If $v = 4$ when $t = 1$ and $s = 6$ when $t = 1$, find s in terms of t.

7. Find the area under the curve $y = 1/(x+1)^2$ from $x = 1$ to $x = 2$.

8. Find the equation of the tangent to $y = 1/(1+x^2)$ at the point where $x = 1$.

9. Find the greatest value of $1/(1+x^2)$.

10. By considering the function $f: x \to \sin(mx+c)$ as a composite function show that its derived function is given by

$$f': x \to m\cos(mx+c).$$

Find the derived function of $g: x \to \cos(mx+c)$. What are:

$$\int^t \cos(mx+c)\,dx \quad \text{and} \quad \int^t \sin(mx+c)\,dx?$$

11. The side of a cube is measured as 4 ± 0.01 cm. If the volume is calculated from this measurement, what approximately is its possible range?

12. A metal sphere is heated and its radius expands from 2 cm to 2·02 cm. What is the approximate change in volume?

13. The radius of a soap bubble changes from 1 cm to 0·9 cm. What is the approximate change in its surface area?

14. Find the approximate value of

$$x^3+2x^2-3x+1,$$

when $x = 1·002.$

15. The circumference of a circular cylinder is measured using a piece of string which can stretch by up to 1 % of its length. Assuming the height is measured correctly, find the approximate percentage error in the volume of the cylinder.

16. The radius of a circular aperture expands from 2 cm to 2·1 cm. Calculate the approximate change in area.

17. A cylinder of an engine has diameter 8 cm. It is made of metal which expands by 0·000012 for every degree C rise in temperature. Find the approximate change in cross-sectional area when the temperature rises from 10 °C to 90 °C.

18. Find the gradient functions of

$$(a) \quad x \to 3x^2 + \frac{4}{x}; \qquad (b) \quad x \to x^3 + 2x - \frac{7}{x}; \qquad (c) \quad x \to \frac{1}{x} + 1.$$

19. At which points does the gradient function of

$$x \to x + 1/x$$

have the values (a) 0, (b) −3, (c) 3?

20. Prove that the rectangle of given area which has least perimeter is a square.

21. An open rectangular tank with a volume of 200 m³ is to be constructed from a sheet of metal. Find the least amount of metal required. (Assume that a square base is used.)

22. Prove that a cylindrical tin of given volume has least surface area when the height is twice the radius.

23. What is the least value of the sum of a positive number and its reciprocal?

14

VECTORS AND THEIR USES

1. THE DEFINITION OF A VECTOR

A vector can either be defined algebraically or geometrically. Algebraically, a two-dimensional vector is simply a column of two numbers such as $\begin{pmatrix} 5 \\ -4 \end{pmatrix}$; geometrically a vector is a number together with a direction, of which the most familiar example is a *translation*, the number giving the length of the translation, and the direction being the direction of the translation. The connection between the two is simple; the vector $\begin{pmatrix} 5 \\ -4 \end{pmatrix}$ is associated with the translation which takes the origin, the point (0, 0), to the point (5, −4); when we associate a vector with a point in this way we call it the *position vector* of the point.

If the point is P, we can write the position vector as **OP**; position vectors all start from O. But since a vector defines a translation of the *whole plane*, the segment traced out by any point completely specifies it and is a suitable name for it: if $\overline{AB}$, $\overline{CD}$, $\overline{EF}$, ... are equal and parallel, then **AB**, **CD**, **EF** are alternative names for the same vector, which could equally well be described, say, as '3 units north-east'.

We can also use single letters to denote vectors; thus we could write

$$\mathbf{a} = \mathbf{OA} = [2, \text{east}] = \begin{pmatrix} 2 \\ 0 \end{pmatrix},$$

and these are all different ways of saying the same thing, which are useful in different contexts.

2. ADDITION AND SUBTRACTION

The sum of two vectors is obtained by choosing two segments which represent the vectors and are placed 'head to tail', that is, one segment starts from where the other leaves off.

In Figure 1 $\overline{AB}$ and $\overline{BC}$ are two such segments, and we define

$$\mathbf{AB} + \mathbf{BC} = \mathbf{AC}.$$

(Note that we put two arrows on the sum vector.) This is called the 'triangle law' of addition.

This is the same as saying that the sum of two vectors is the vector corresponding to the translation which is equivalent to the two translations performed in succession. Since it does not matter in which order they are performed, we may write either **AB**+**BC** or **BC**+**AB**. In the second case the segments are not 'head to tail', but if we choose **CD** as an alternative

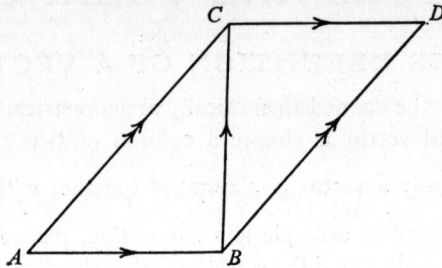

Fig. 1

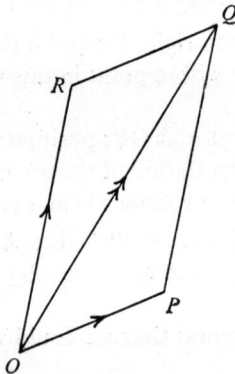

Fig. 2

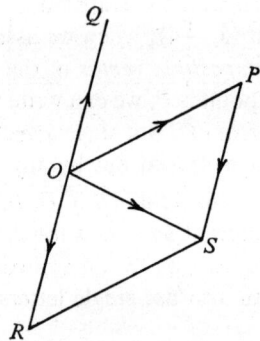

Fig. 3

representation of **AB** then **BD** represents the sum vector, and obviously $\overline{BD}$ and $\overline{AC}$ are equal and parallel. If we take *position vectors* starting from O (see Figure 2), then if $OPQR$ is a parallelogram,

$$\mathbf{OP}+\mathbf{OR} = \mathbf{OP}+\mathbf{PQ} = \mathbf{OQ},$$

a rule which is called the *parallelogram law* for addition of position vectors.

The zero vector is the vector of zero length—that is, the identity translation, and the position vector of O.

The *negative* of a given vector is the vector which when added to it gives the zero vector; that is, it has the same magnitude in the opposite direction. Thus $-\mathbf{AB} = \mathbf{BA}$, and $\mathbf{AB}+\mathbf{BA} = \mathbf{0}$.

To *subtract* one vector from another is to add its *negative*; thus

$$\mathbf{OP}-\mathbf{OQ} = \mathbf{OP}+\mathbf{QO}$$
$$= \mathbf{OP}+\mathbf{OR}$$
$$= \mathbf{OS} \quad \text{(see Figure 3)}.$$

It is obvious, since $\overline{RO} = \overline{OQ} = \overline{SP}$, that the triangle OQP is a translation of the triangle ROS, so that $OSPQ$ is a parallelogram and $\mathbf{OS+OQ = OP}$, which coincides with our intuitive ideas about subtraction.

Alternatively, we can write $\mathbf{OP-OQ = QP}$.

Exercise A

1. Construct line-segments on your paper, marked with arrows to represent the following vectors:

$$\mathbf{a}\ (2\ \text{units, east}); \quad \mathbf{b}\ (5\ \text{units, north-west}); \quad \mathbf{c}\ (4\ \text{units, south}).$$

2. With the data of Question 1, construct the vectors:

$$\mathbf{a+b}, \quad \mathbf{(a+b)+c}, \quad \mathbf{b+c}, \quad \mathbf{a+(b+c)}.$$

Does $\qquad\qquad\qquad \mathbf{(a+b)+c = a+(b+c)}$?

3. With the data of Question 1, construct the vectors $\mathbf{a-b}$, $\mathbf{b-c}$.

4. If $\mathbf{OA = a}$, $\mathbf{OB = b}$, what is $\mathbf{b-a}$?

If $\mathbf{OC = c}$, $\mathbf{OD = d}$, and $\mathbf{b-a = c-d}$, what can you say about the figure $ABCD$?

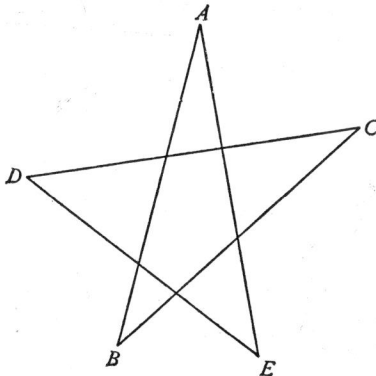

Fig. 4

5. Prove that, in Figure 4, $\mathbf{AB+BC+CD+DE+EA = 0}$.

A point P is chosen and

$$\left.\begin{array}{c} \overline{PQ} \\ \overline{QR} \\ \overline{RS} \\ \overline{ST} \\ \overline{TU} \end{array}\right\} \text{is drawn equal and parallel to} \left\{\begin{array}{c} \overline{AB} \\ \overline{DE} \\ \overline{BC} \\ \overline{EA} \\ \overline{CD}, \end{array}\right.$$

so making a convex figure $PQRSTU$. What can you say about $\mathbf{UP}$, and why?

6. *ABCDEF* is a regular hexagon. If **AB** = **a**, **BC** = **b**, **CD** = **c**, express in terms of **a**, **b**, **c** the vectors represented by

(*a*) **DE**;	(*b*) **EF**;	(*c*) **FA**;	(*d*) **AD**;
(*e*) **BE**;	(*f*) **CE**;	(*g*) **AE**;	(*h*) **CA**.

Would your answers be the same if:
  (i) the hexagon was not regular,
  (ii) the hexagon was not convex,
  (iii) the hexagon was not all in one plane?

3. TRACK AND COURSE

The addition of vectors is helpful in solving problems of navigation for planes or boats when the medium, air or water, in which they are moving, is itself in motion.

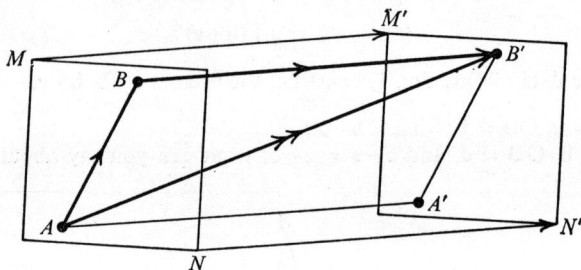

Fig. 5

Suppose a square of paper, with opposite corners *M* and *N*, is translated steadily and uniformly over the table top to the position *M′ N′*. *A′* and *B′* are the new positions of two points *A* and *B*. Hence

$$\mathbf{AA'} = \mathbf{BB'} = \mathbf{MM'} = \mathbf{p},$$

say. Suppose also that while the paper is moving a spider walks uniformly along the line *AB*, so that at the beginning of the interval the spider was at *A*, and at the end of the interval it was at *B′*, where

$$\mathbf{AB} = \mathbf{A'B'} = \mathbf{q}.$$

Relative to the table the spider has moved from *A* to *B′*, along the line *AB′*, and **AB′** = **p**+**q**.

If these movements took place in unit time, then the vector **p** would represent the velocity of the paper, **q** the velocity of the spider relative to the paper, and **p**+**q** the resultant velocity of the spider.

If *A* and *B* (Figure 6) were floating pieces of driftwood on the surface of water in a tidal estuary, and a boat moved from *A* to *B* while the water moved bodily so that *A* moved to *A′* and *B* to *B′*, then the actual movement of the boat, relative to the land, would be represented by **AB′**.

For an aeroplane, we could imagine *A* and *B* to be two cloud forms in a large mass of air moving steadily over the earth, the movement of the air being given by **AA'**. In this case **AB** represents the velocity of the plane *relative* to the air (see Figure 7). The magnitude of $\overline{AB}$ represents the *air speed*, and the direction of $\overrightarrow{AB}$ is the direction of attempted flight, the *course*. **BB'** represents the *wind velocity* (which has been exaggerated in relation to a typical aeroplane's speed). An observer on the ground would observe the plane moving along the line $\overrightarrow{AB'}$; the *track* $\overline{AB'}$, on the same scale as $\overline{AB}$, represents the *ground speed*. The angle BAB' is called the *drift*.

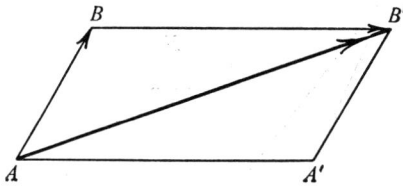

Fig. 6

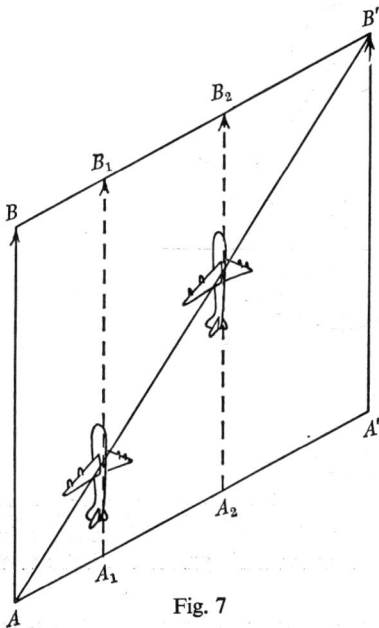

Fig. 7

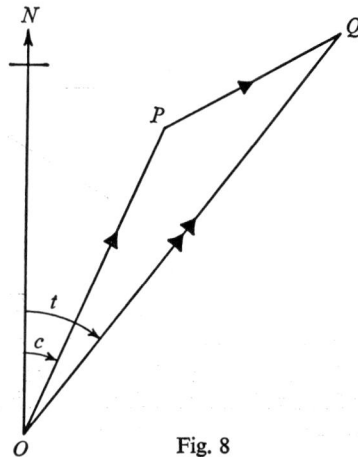

Fig. 8

Summarizing (see Figure 8),

    OP represents the plane's air speed (length) and course (direction).

    PQ represents wind speed and direction.

    OQ represents the plane's ground speed (length) and track (direction).

Example 1. A plane is to fly, at an air speed of 250 km/h, from Oxford to Birmingham 100 km away on a bearing of 335°. The wind is blowing *from*

240° at 50 km/h. Find the course and ground speed and the time of the flight.

Draw OB in the direction of the track (see Figure 9). OR is the wind speed on a suitable scale. With centre R and radius equal to the air speed on the same scale, draw an arc to cut OB at Q. Then $\overline{OQ}$ is the ground speed, and α is the course which the pilot sets.

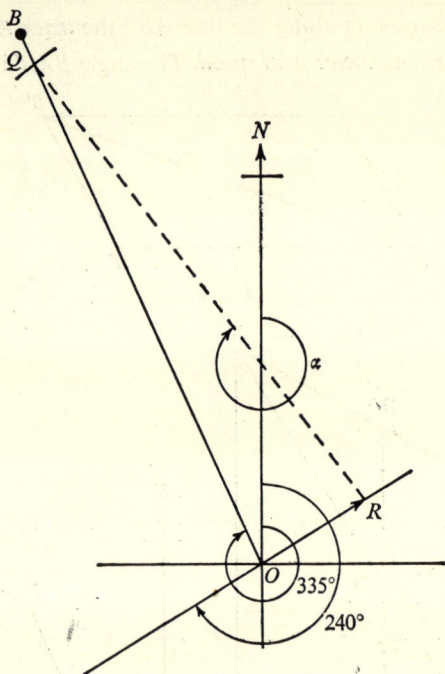

Fig. 9

Exercise B

1. In a river flowing due south at 5 km/h between straight parallel banks, a man swims at 3 km/h so that his body is at right-angles to the line of flow. If the river is 30 m wide how much is he carried down in swimming from one bank to the other?

2. A small plane is heading due north (that is, the line of the fuselage is NS) with an air speed of 150 km/h. If there is a SW wind of 60 km/h, in what direction does the plane actually move (the course) and with what speed relative to the ground (the 'ground speed')? (A SW wind blows *from* the SW.)

3. A boat has a speed of 10 knots and is to travel from A to B. The tide is setting at 3 knots in the direction shown by AC (Figure 10). Find, by drawing, the direction in which the boat must be steered to travel along AB. What is its effective speed?

4. A vector of 3 units due east is added to one of 4 units due north. In what direction should you add a third vector of magnitude 12 units to give a result equivalent to a single vector:

   (a) due north;   (b) of magnitude 13 units?

5. Assuming a wind speed of 50 km/h from the west, and an air speed of 300 km/h find the course to be set and the ground speed when the track made good is to be on a bearing of:

   (a) 036°;   (b) 216°;   (c) 324°.

How long will it take to fly from Exeter to Lincoln (350 km on a bearing of 036°), and how long will the return journey take?

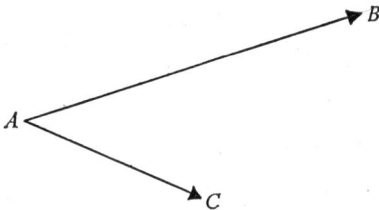

Fig. 10

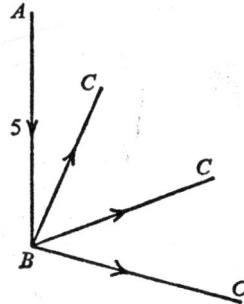

Fig. 11

Problems involving least angles

6. Refer back to the data of Question 1. Draw a line AB 5 units long to represent the velocity of the river (see Figure 11). The vector **BC**, 3 units long, must be added to this to give **AC**, the actual velocity. What is the set of possible positions of C? Which will give a velocity nearest to straight across the river? What is the least distance the swimmer need be carried down stream, and in what direction must he swim to achieve this?

7. A destroyer capable of 30 knots sights an M.T.B. due east at a distance of 5 nautical miles. The M.T.B. is travelling due north at 38 knots. Assuming adequate seaway, how near can the destroyer approach the M.T.B. if it keeps on its course, and in what direction should she steer to do so?

   (Suppose the M.T.B. is reduced to rest by a tide running south at 38 knots affecting the whole sea. In which direction will the destroyer move relative to the M.T.B.? Consider her various tracks in the tideway and find which takes her nearest to the M.T.B. Compare Question 6 and solve by drawing.)

4. FORCES

4.1 Triangle of forces. An important application of the idea of vector addition is the physical relationship between forces acting on a body. A typical experiment is shown in Figure 12.

Two weights, B and C, are suspended from a cord passing over two pulleys. From a knot in the cord at O a third weight, A, hangs freely. It is found that if a triangle PQR is drawn, with sides parallel to the three strings meeting at O, then the lengths of these sides are proportional to the corresponding weights.

293

This experiment is interpreted as follows. The weights exert *forces* on the strings which support them. These forces are equal to the earth's pull on the weights. In strict scientific language we reserve the word 'weight' to describe such a pull; that is, *weight* is a force, the force with which the earth attracts a body.

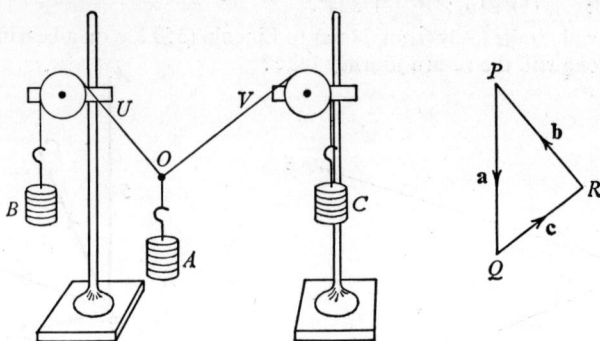

Fig. 12

Now forces are measured by the accelerations they produce. A full discussion of this must be reserved for Chapter 15, but it is enough to say here that a force which gives a mass of 1 kg an acceleration of 1 m/s² is called a *newton*. Since the earth attracts a mass of 1 kg with such a force which, uninhibited—that is, if we drop the mass—will give it an acceleration of about 9·8 m/s², we can say that the weight of 1 kg is about 9·8 *newtons*.

The weight of 1 kg is 9·8 N.

In our case, we assume that the strings passing over the pulleys transfer forces without loss; at O, therefore, three forces are acting, whose magnitudes are the weights of A, B, C, and whose directions are the directions of the three strings at O. The experiment shows that the *vector sum* of these three forces is zero.

This result is usually known as the '*triangle of forces*'; it can be extended to any number of forces and stated in the following form:

> If a number of forces act on a particle (a body small enough to be considered as located at one point) and keep it in equilibrium, their vector sum is zero.

Example 2. In Figure 12, if $A = 10$ kg, $B = 8$ kg, and the strings OU, OV are at right-angles, then in the triangle PQR, **a** $(= $ **PQ**$)$ represents the

294

force in OA, $\mathbf{b}$ ($= \mathbf{RP}$) the force in OB, and $\mathbf{c}$ ($= \mathbf{QR}$) the force in OC; $\mathbf{a}+\mathbf{b}+\mathbf{c} = \mathbf{0}$. Since $PQ = 10 \times 9 \cdot 8$, $PR = 8 \times 9 \cdot 8$ and R is a right-angle, we have at once that $QR = 6 \times 9 \cdot 8$ and the angle $P = 36 \cdot 9°$. Hence the force of tension in $OV = 6 \times 9 \cdot 8$ kg and C must be 6 kg. Further the strings OU, OV are inclined to the vertical at $36 \cdot 9°$ and $53 \cdot 1°$.

4.2   Contact forces. If one body is in contact with another, there is usually a force between them. A very large force can be called into play to resist penetration or deformation of one body by the other; a much smaller force is available to resist relative motion.

> The force that body A exerts on body B is *always* equal and opposite to the force that body B exerts on body A, whether the bodies are at rest, moving together, or in relative motion.

This is the Law of Reaction (Newton's Third Law) and enables us to connect actions on different bodies. Mathematically, if $\mathbf{P}$ is the force A exerts on B, then the force B exerts on A is $-\mathbf{P}$.

If $\mathbf{P}$ is perpendicular (normal) to the surface of contact between A and B, we say that the contact is *smooth*, since there is no force to resist sideways motion. We then call $\mathbf{P}$ the *normal action* of A on B. If the contact is not smooth, so that $\mathbf{P}$ is not normal to the surface, we often regard it as the sum of two vectors, $\mathbf{F}$ and $\mathbf{N}$, where $\mathbf{F}$ (the *friction*) lies in the surface, and $\mathbf{N}$ (the normal action) is at right-angles to it.

Example 3. A car weighing 1000 kg is at rest on a slope of $10°$ with its brakes on. Describe the forces on it.

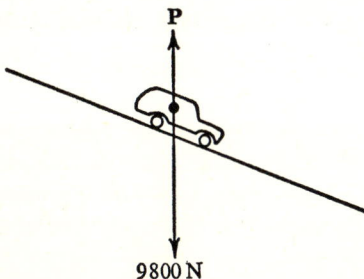

9800 N

Fig. 13

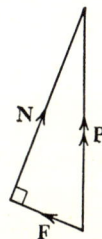

Fig. 14

The weight is a vertical force of 9800 kg. Since the car is at rest, the vector sum of this and the action of the ground on the car is zero; hence the action of the ground, $\mathbf{P}$, is a vertical force of 9800 kg. If we like to divide (artificially) $\mathbf{P}$ into two vectors $\mathbf{N}$ and $\mathbf{F}$, where $\mathbf{P} = \mathbf{N}+\mathbf{F}$, we see from Figure 14 that $\mathbf{F}$ is up the plane, of magnitude

$$9800 \sin 10° = 1700 \text{ N},$$

and **N** is of magnitude
$$9800 \cos 10° = 9600 \text{ N}.$$
Since the tyres are called upon to provide a frictional force equal to a weight of over 150 kg it would be as well to put stones under the wheels.

Exercise C

1. In Figure 12, if the weights A, B, and C are 4 kg, 3·6 kg and 2·5 kg respectively, draw the vector diagram and find the expected inclination of the strings to the vertical (by drawing and measurement).

2. If the weight B is 12 kg, and C is 5 kg, what weight A will pull the string down until angle UOV is a right-angle? (See Figure 12.)

3. A truck on freely-turning wheels is suspended on an incline by a counterpoise as shown in Figure 15. The truck weighs 2000 kg and the incline is 1 in 10 (that is, the sine of the angle is 1/10). What counterpoise is needed?

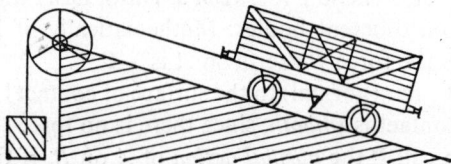

Fig. 15

4. A picture weighing 5 kg hangs by a cord passing over a hook, both parts of which are inclined at 40° to the horizontal. Draw the vector triangle for the forces on the hook, and find the tension in the string.

5. An aircraft is in straight level flight with constant speed (in these circumstances the forces on it balance). Its propellers develop a forward thrust of 230000 newtons and the aircraft weighs 50000 kg. Find the magnitude and direction of the force of the air on it. (See Figure 16.)

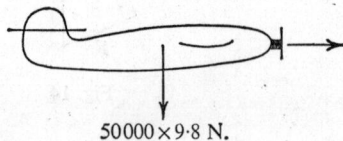

$50000 \times 9·8$ N.

Fig. 16

10^5 N   45° 8×10^4 N

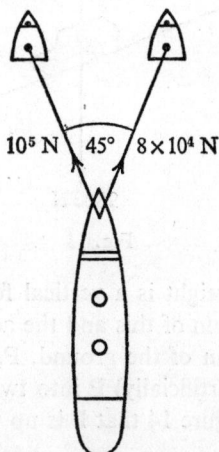

Fig. 17

6. Two tugs are pulling a liner with forces (assumed horizontal) of 10^5 N and 8×10^4 N; the angle between their hawsers is 45°. (See Figure 17.) In what direction will the liner move, and what is the resistance of the water in steady motion?

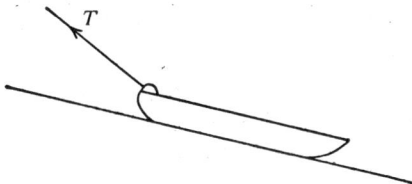

Fig. 18

7. A toboggan is being pulled up a slope inclined at 15° to the horizontal by a rope inclined at 25° to the slope (see Figure 18). The toboggan weighs 9 kg and the friction is 25 N. Draw a diagram with all the forces clearly marked, then draw a vector diagram and find T, the pull in the rope.

5. MULTIPLICATION OF A VECTOR
BY A NUMBER

To multiply a vector by a number k, we simply multiply its magnitude by k and leave its direction unchanged. This is the same (if k is an integer) as adding k vectors, all equal, end to end in the same direction.

We ought to check that expressions like 2(**AB**+**BC**) and 2**AB**+2**BC** will mean the same thing. Look at Figure 19.

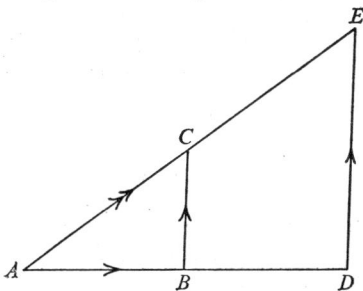

Fig. 19

AB+**BC** = **AC**, so to get 2(**AB**+**BC**) we double AC and continue the line $\overline{AC}$ to $\overline{AE}$. 2**AB** is **AD**, where $\overline{AD} = 2\overline{AB}$. It only remains to verify that **DE** = 2**BC**, and this is true. (Why?)

Hence, **AD**+**DE** = **AE** gives us

$$2\mathbf{AB}+2\mathbf{BC} = 2\mathbf{AC} = 2(\mathbf{AB}+\mathbf{BC}).$$

Multiplication by 2 is therefore distributive over addition, and general results can be proved in the same way.

5.1 Mid-point. A useful result is as follows:

If M is the mid-point of $\overline{PQ}$, and O is any other point, then

$$OP+OQ = 2OM.$$

The proof is simple. From Figure 20 we have

$$OP+OQ = OM+MP+OM+MQ \quad \text{(triangle law)}$$

$$= 2OM+MP+MQ$$

$$= 2OM$$

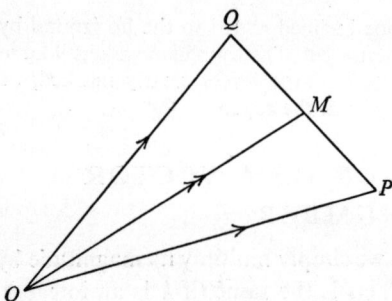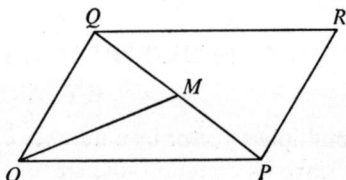

| Fig. 20 | Fig. 21 |

since $\mathbf{MP} = -\mathbf{MQ}$, given that M is the mid-point of PQ.

In Figure 21 $OPRQ$ is a parallelogram, and we know that

$$OP+OQ = OR \quad \text{and} \quad OP+OQ = 2OM.$$

What does this tell us about M?

Exercise D

1. Suppose $\mathbf{p}$ is a vector of magnitude 3 units east; $\mathbf{q}$ is 2 units SW, and $\mathbf{r}$ is 1 unit in a direction N 15° W (345°). Take north at the top of your paper and construct line-segments to represent

(a) $3\mathbf{q}$;   (b) $-2\mathbf{r}$;   (c) $\mathbf{p}+\tfrac{1}{2}\mathbf{q}$;

(d) $2\mathbf{p}+4\mathbf{q}+5\mathbf{r}$;   (e) $\mathbf{q}-3\mathbf{r}$.

2. Take a point O and let $\mathbf{OP}$ represent $\mathbf{p}$ and $\mathbf{OR}$ represent $\mathbf{r}$, as given in Question 1. Find the point Q if $3OQ = OP+2OR$. What do you notice about PQR?

3. Figure 22 shows a cube; the vectors **OA**, **OB**, **OC** are called **i**, **j** and **k**. Describe the location of the points whose position vectors (that is, vectors starting out from O) are

(a) $\mathbf{i}+\mathbf{j}$;   (b) $\mathbf{i}+\mathbf{j}+\mathbf{k}$;   (c) $\tfrac{1}{2}\mathbf{i}+\tfrac{1}{2}\mathbf{j}$;

(d) $\mathbf{k}+\tfrac{1}{2}\mathbf{j}$;   (e) $\tfrac{1}{2}(\mathbf{i}+\mathbf{j}+\mathbf{k})$.

4. With the data of Question 3, give the position vectors (in terms of **i, j, k**) of (*a*) *F*; (*b*) *E*; (*c*) the mid-point of $\overline{AC}$; (*d*) the mid-point of $\overline{FG}$; (*e*) the centre of the face *FCEG*.

5. *O* is any point, and *P, Q, R* are the mid-points of the sides of a triangle *ABC*. Prove that **OP + OQ + OR = OA + OB + OC**.
    [Use the mid-point theorem three times and add.]

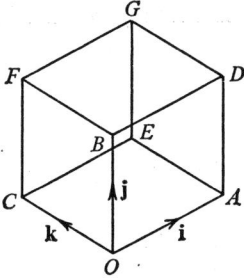
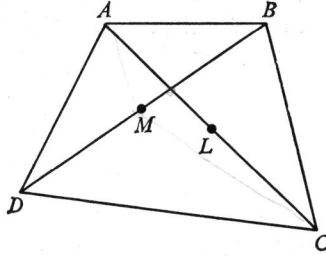

Fig. 22   Fig. 23

6. In Figure 23 *ABCD* is a quadrilateral and *L, M* are the mid-points of $\overline{AC}$, $\overline{BD}$. Prove that
$$AB + AD + CB + CD = 4LM.$$

[Use the mid-point theorem twice.]
What does **AB + BC + CD + DA** equal? Hence show that **AB + CD = 2LM**.
[Add.]

7. Deduce from this a theorem about quadrilaterals for which *L* and *M* coincide.

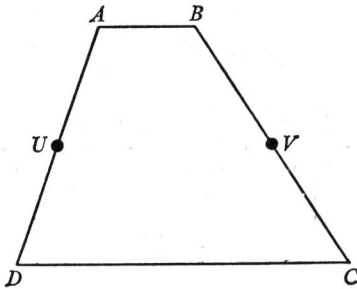

Fig. 24

8. In Figure 24, *ABCD* is a trapezium; *AB* is parallel to *DC* and **DC = 3AB**. *U* is the mid-point of $\overline{AD}$ and *V* is the mid-point of $\overline{BC}$. Prove **UV = 2AB**. [Use Question 6.]

6. COMPONENTS

If we take an origin and coordinate axes in the plane, then if the co-ordinates of a point P are (a, b), the vector **OP** is associated with the translation from O to P which we have written as the column vector $\begin{pmatrix} a \\ b \end{pmatrix}$.

The two numbers a and b are called the *components* of the vector **OP** along the axes. Thus **OP** is 'a along and b up', or strictly, 'a units in the x-direction and b units in the y-direction'.

The vectors

$$\begin{pmatrix} 1 \\ 0 \end{pmatrix} \quad \text{and} \quad \begin{pmatrix} 0 \\ 1 \end{pmatrix}$$

we have already met as *unit-vectors* or *base-vectors* in the direction of the axes. We write them usually with the single letters **i** and **j**. (This is of course connected with the j of complex numbers but must not be confused with it; this **j** is a vector, not a number.) Since

$$\begin{pmatrix} a \\ b \end{pmatrix} = a\begin{pmatrix} 1 \\ 0 \end{pmatrix} + b\begin{pmatrix} 0 \\ 1 \end{pmatrix}$$

$$= a\mathbf{i} + b\mathbf{j},$$

we can say either   '**OP** has components a and b'

or   '**OP** $= a\mathbf{i} + b\mathbf{j}$',

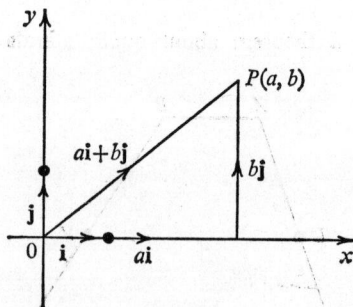

Fig. 25

and these mean the same thing. The coordinates of P are then (a, b).

In the same way we can speak of the components of a force or a velocity. We have already seen that it can be helpful to consider a force **P** as made up of a normal action **N** perpendicular to the plane of contact, and a friction **F**, in this plane (see Figure 26). F and N are then the *components of P* in these directions, and

$$\mathbf{P} = \mathbf{F} + \mathbf{N}.$$

300

We can also find the components of a velocity. If we throw a ball up at 60° to the horizontal with a speed of 20 m/s (see Figure 27), then the horizontal and vertical components of this velocity are given by ON and NP, and they are 20 cos 60° or 10 m/s and 20 sin 60° or 17 m/s approximately. We shall find these useful in discovering the path of the ball after it is thrown.

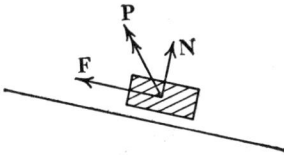
Fig. 26

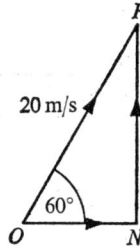
Fig. 27

Exercise E

1. If P is (a, b) and Q is (c, d), write **OP** and **OQ** in terms of **i** and **j**. What are the components of

   (a) **OP** + **OQ**;   (b) **OP** − **OQ**;   (c) k**OP**?

If **OP** + **OQ** = **OR** and **OP** − **OQ** = **OS**, what are the coordinates of R and S? Draw a diagram to show clearly the positions of

$$P, Q, R, S \quad \text{when} \quad a = 4, b = 1, c = -2, d = 5.$$

2. If M is the mid-point of $\overline{PQ}$, and P is $(2, -1)$ and Q is $(4, 7)$:
   (a) express **OP** and **OQ** in terms of **i** and **j**;
   (b) express **OM** in terms of **i** and **j** (use Section 5);
   (c) give the coordinates of M.

3. Repeat Question 2 if P is $(-5, 3)$ and Q is $(6, -1)$.

4. Repeat Question 2 if P is (x_1, y_1) and Q is (x_2, y_2). In this way get a formula for the coordinates of the mid-point of $\overline{PQ}$.

5. Use your formula to find the mid-point of $\overline{PQ}$ when P is $(-1·6, 2·3)$ and Q is $(0·5, 4·1)$.

6. $PQRS$ is a parallelogram. P is $(2, 1)$, Q is $(-1, 5)$, S is $(0, -3)$. Find the coordinates of R and of the point where PR meets QS.

7. If I travel 10 km on a bearing of 050°, what are the eastward and northward components of my displacement?

8. Answer Question 7 if the bearing is 295°.

9. An aeroplane is climbing at 400 km/h at an inclination of 10° to the horizontal. What is the vertical component of its velocity?

10. I push a lawn mower with a force of 150 N directed down the handle, at 55° to the horizontal. What is the horizontal component of my push?

If the weight of the mower is 10 kgf, what is the normal action of the ground on it? (The vertical forces must balance.)

11. Looked at horizontally, a bird appears to be climbing at 5 m/s; viewed from vertically above, it appears to be covering the ground at a speed of 10 m/s. How fast is it actually flying?

12. Two men are pushing a car with forces of 150 N and 200 N at angles of 10° and 16° with the fore-and-aft line. What is their combined effective forward push?

7. SOME EXPRESSIONS IN COORDINATES

This section is merely a stocktaking of some familiar and useful results in the language of coordinates.

7.1 Distance. If O is the origin and P is the point (x, y), then by Pythagoras's theorem the distance

$$OP = \sqrt{(x^2+y^2)}.$$

The length of the vector

$$\begin{pmatrix} x \\ y \end{pmatrix} = \mathbf{OP} = x\mathbf{i}+y\mathbf{j}$$

is thus always $\sqrt{(x^2+y^2)}$.

The vector from (x_1, y_1) to (x_2, y_2) is

$$\begin{pmatrix} x_2-x_1 \\ y_2-y_1 \end{pmatrix};$$

this distance is therefore

$$\sqrt{\{(x_2-x_1)^2+(y_2-y_1)^2\}}.$$

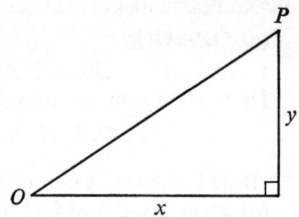

Fig. 28

7.2 Gradient. The gradient of the same vector **OP** is defined as y/x; it is the tangent of the angle from the x-axis to **OP**.

The gradient of the vector from (x_1, y_1) to (x_2, y_2) is

$$\frac{y_2-y_1}{x_2-x_1}.$$

If this is m, then $y_2-y_1 = m(x_2-x_1)$,

and the point (x_2, y_2) belongs to the set

$$\{(x, y): y-y_1 = m(x-x_1)\}$$

which is the equation of the line through (x_1, y_1) with a gradient m.

302

7.3   Sine and cosine. The general definition of $\sin \theta$ and $\cos \theta$ is as follows (see Figure 29):

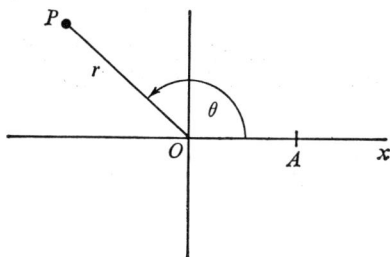

Fig. 29

Let the central direction OA be taken as the x-axis, let $\widehat{AOP} = \theta$ be a counter-clockwise rotation, and let OP be of length r.

Then
$$\cos \theta = \frac{\text{the } x\text{-component of } \mathbf{OP}}{r},$$

and
$$\sin \theta = \frac{\text{the } y\text{-component of } \mathbf{OP}}{r}.$$

That is to say that if P is (x, y) then
$$x = r \cos \theta \quad \text{and} \quad y = r \sin \theta.$$

7.4   Mid-point. From the vector equation in Section 5
$$\mathbf{OP} + \mathbf{OQ} = 2\mathbf{OM},$$
where M is the mid-point of $\overline{PQ}$, we can deduce at once the coordinates of M in terms of those of P and Q. Suppose P is (x_1, y_1) and Q (x_2, y_2). Then
$$\mathbf{OP} + \mathbf{OQ} = x_1\mathbf{i} + y_1\mathbf{j} + x_2\mathbf{i} + y_2\mathbf{j}$$
$$= (x_1 + x_2)\mathbf{i} + (y_1 + y_2)\mathbf{j} = 2\mathbf{OM}.$$

Hence M is the point $\left(\dfrac{x_1 + x_2}{2}, \dfrac{y_1 + y_2}{2}\right).$

Exercise F

For the following pairs of points, write down:
   (a) the length of PQ;
   (b) the gradient of PQ;
   (c) the coordinates of the mid-point of $\overline{PQ}$;
   (d) the sine and cosine of the angle from Ox to $\mathbf{PQ}$.

1. $P(2, -3)$; $Q(5, 1)$.   2. $P(-4, 7)$; $Q(8, 2)$.   3. $P(5, -3)$; $Q(-1, 5)$.

4. Find the mid-points P, Q, R, S of the sides AB, BC, CD, DA of the quadrilateral $ABCD$, where A is $(-1, 2)$, B is $(3, 0)$, C is $(5, 3)$, and D is $(0, 9)$. Find the lengths and gradients of PQ and SR. What conclusion do you draw?

5. Show that the result of Question 4 does not depend on the coordinates of *A, B, C, D,* as follows.

$$\left.\begin{array}{l} \mathbf{OA}+\mathbf{OB} = 2\mathbf{OP} \\ \mathbf{OB}+\mathbf{OC} = ... \end{array}\right\} \Rightarrow 2\mathbf{PQ} = 2(\mathbf{OQ}-\mathbf{OP}) = ...$$

Write down two similar equations for 2**OS** and 2**OR** and use them to find 2**SR**. Hence prove that **SR** = **PQ**. What is the conclusion?

8. THE SCALAR PRODUCT

8.1 Definition. If **p** and **q** are two vectors, of lengths *p* and *q*, and θ is the angle between them, we define their *scalar product* (sometimes called the *dot product*) **p.q** as the *number pq* cos θ.

Since cos θ = cos $(360° - \theta)$ it makes no difference which way we rotate from **p** to **q**, nor even if we rotate from **q** to **p** instead; in fact

$$\mathbf{p.q} = \mathbf{q.p} = pq \cos \theta.$$

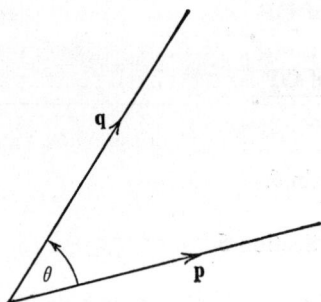

Fig. 30

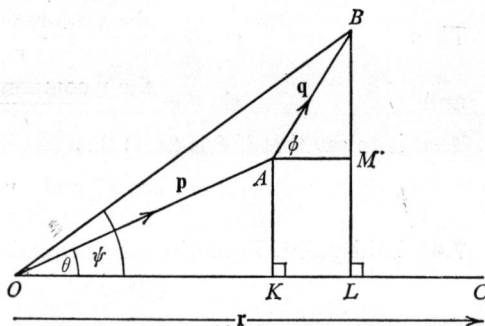

Fig. 31

If this is a reasonable sort of product, we must be sure that it satisfies the usual laws for multiplication; the commutative law we have already shown to be satisfied; the associative law does not apply, since **p.q** is a number or scalar, and cannot be an element in a further scalar product. There remains the distributive law.

$$(\mathbf{p}+\mathbf{q}).\mathbf{r} = \mathbf{p.r}+\mathbf{q.r}.$$

To establish this, we consider Figure 31 in which **OA** = **p**, **AB** = **q**, **OC** = **r**. Then **p**+**q** = **OB**, and if $\widehat{COA} = \theta$, $\widehat{MAB} = \phi$ and $\widehat{COB} = \psi$, we have

$$\mathbf{p.r}+\mathbf{q.r} = OA.OC \cos \theta + AB.OC \cos \phi$$

$$= OC.OK+OC.AM = OC.(OK+KL)$$

$$= OC.OL = OC.OB \cos \psi = (\mathbf{p}+\mathbf{q}).\mathbf{r},$$

which establishes the result.

304

8.2   Applications. Two of the most important uses of the scalar product are as follows:

(1) $\mathbf{p} \cdot \mathbf{p} = p^2$, the square of the length of p, since $\cos 0° = 1$.

(2) If $\mathbf{p}$ and $\mathbf{q}$ are perpendicular, since $\cos 90° = 0$, it follows that $\mathbf{p} \cdot \mathbf{q} = 0$, and conversely, if $\mathbf{p} \cdot \mathbf{q} = 0$, and $\mathbf{p}$ and $\mathbf{q}$ are neither of them zero, then $\cos \theta = 0$ and $\mathbf{p}$ is perpendicular to $\mathbf{q}$.

Pythagoras's theorem.

If $\mathbf{i}, \mathbf{j}$ are unit vectors along the perpendicular axes, then

$$\mathbf{i} \cdot \mathbf{i} = \mathbf{j} \cdot \mathbf{j} = 1 \quad \text{and} \quad \mathbf{i} \cdot \mathbf{j} = \mathbf{j} \cdot \mathbf{i} = 0.$$

Hence if $\mathbf{p} = \mathbf{OP} = x\mathbf{i} + y\mathbf{j}$ is any vector,

$$p^2 = \mathbf{p} \cdot \mathbf{p} = (x\mathbf{i} + y\mathbf{j}) \cdot (x\mathbf{i} + y\mathbf{j})$$
$$= x^2 \mathbf{i} \cdot \mathbf{i} + 2xy \mathbf{i} \cdot \mathbf{j} + y^2 \mathbf{j} \cdot \mathbf{j}$$
$$= x^2 + y^2,$$

or $OP^2 = OL^2 + OP^2,$

which is the well-known result ascribed to Pythagoras.

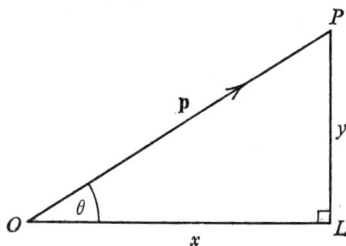

Fig. 32

Components

x and y in Figure 32 are called the *components* of $\mathbf{p}$; plainly

$$x = \mathbf{p} \cdot \mathbf{i} = p \cos \theta, \quad \text{and} \quad y = \mathbf{p} \cdot \mathbf{j} = p \cos (90° - \theta) = p \sin \theta.$$

Further, the general definitions of $\sin \theta$ and $\cos \theta$ given in Section 7.3 have been chosen to make these results true for all angles θ.

$$\mathbf{p} \cdot \mathbf{i} = x = p \cos \theta,$$
$$\mathbf{p} \cdot \mathbf{j} = y = p \sin \theta,$$
$$\mathbf{p} = \begin{pmatrix} x \\ y \end{pmatrix} = x\mathbf{i} + y\mathbf{j}.$$

Also, since $x^2 + y^2 = p^2$, always we have

$$\cos^2 \theta + \sin^2 \theta = 1.$$

8.3   General scalar product in terms of components. If

$$\mathbf{p}_1 = x_1\mathbf{i}+y_1\mathbf{j} \quad \text{and} \quad \mathbf{p}_2 = x_2\mathbf{i}+y_2\mathbf{j},$$

then
$$\mathbf{p}_1\cdot\mathbf{p}_2 = (x_1\mathbf{i}+y_1\mathbf{j})\cdot(x_2\mathbf{i}+y_2\mathbf{j})$$
$$= x_1x_2+y_1y_2 \quad \text{(since } \mathbf{i}\cdot\mathbf{j} = 0\text{)}.$$

Now suppose that $\mathbf{p}_1$ and $\mathbf{p}_2$ make angles θ_1 and θ_2 with OX.

Then
$$\mathbf{p}_1\cdot\mathbf{p}_2 = p_1 p_2 \cos(\theta_2-\theta_1)$$
$$= p_1 p_2 \cos(\theta_1-\theta_2);$$

and
$$x_1x_2+y_1y_2 = p_1\cos\theta_1\cdot p_2\cos\theta_2+p_1\sin\theta_1\cdot p_2\sin\theta_2$$

by the definition of cosine and sine.

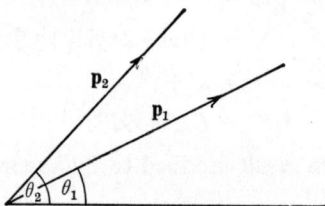

Fig. 33

The equation
$$\mathbf{p}_1\cdot\mathbf{p}_2 = x_1x_2+y_1y_2$$

therefore gives at once (on division by $p_1 p_2$)

$$\cos(\theta_1-\theta_2) = \cos\theta_1\cos\theta_2+\sin\theta_1\sin\theta_2.$$

This result is one of the so-called *addition formulae* which will be discussed more fully in a later section.

Exercise G

1. Find the lengths of the vectors

$$\begin{pmatrix} 4 \\ -3 \end{pmatrix} \quad \text{and} \quad \begin{pmatrix} -24 \\ 7 \end{pmatrix}$$

and the angle between them.

2. Show that the vectors

$$\begin{pmatrix} 2a \\ -5a \end{pmatrix} \quad \text{and} \quad \begin{pmatrix} 5b \\ 2b \end{pmatrix}$$

are perpendicular.

3. Evaluate the products:

(a) $(4\mathbf{i}-\mathbf{j})\cdot(3\mathbf{i}+2\mathbf{j})$;

(b) $(8\mathbf{i}-15\mathbf{j})\cdot(8\mathbf{i}-15\mathbf{j})$;

(c) $(\mathbf{i}+3\mathbf{j})\cdot(\mathbf{i}-3\mathbf{j})$.

306

4. A ladder, 10 m long, rests against a building with its foot 3 m from the base of the building. It is pulled over to rest over the top of a wall 5 m high and 5 m away from the building.

(*a*) How high up the building does it reach?

(*b*) What length of ladder extends beyond the top of the wall?

(*c*) Through what angle has it been turned?

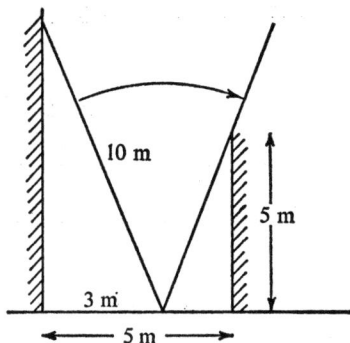

Fig. 34

9. PERPENDICULARITY

9.1   Perpendicular vectors. We have seen that two vectors are perpendicular if their scalar product is zero; that is,

$$\mathbf{p_1} = \begin{pmatrix} x_1 \\ y_1 \end{pmatrix} \quad \text{and} \quad \mathbf{p_2} = \begin{pmatrix} x_2 \\ y_2 \end{pmatrix}$$

are perpendicular if $\quad \mathbf{p_1} \cdot \mathbf{p_2} = x_1 x_2 + y_1 y_2 = 0.$

For example,

$$\begin{pmatrix} 5 \\ -2 \end{pmatrix} \quad \text{is perpendicular to} \quad \begin{pmatrix} 2 \\ 5 \end{pmatrix} \quad \text{and so is} \quad \begin{pmatrix} 5k \\ -2k \end{pmatrix}$$

for any k. This in fact is often the easiest way of handling perpendicular vectors:

$$\text{Any vector perpendicular to} \quad \begin{pmatrix} a \\ b \end{pmatrix} \quad \text{is} \quad \begin{pmatrix} kb \\ -ka \end{pmatrix}.$$

The gradients of these vectors are $\dfrac{b}{a}$ and $\dfrac{-ka}{kb}$, that is, $\dfrac{b}{a}$ and $\dfrac{-a}{b}$, whose product is -1.

Hence:

> If two vectors are perpendicular,
> the product of their gradients is -1.

There is an exceptional case, namely the base vectors themselves.

$$\begin{pmatrix} 1 \\ 0 \end{pmatrix} \quad \text{and} \quad \begin{pmatrix} 0 \\ 1 \end{pmatrix} \quad \text{are perpendicular}$$

by definition, but the second does not have a gradient.

The converse is, however, true without exception; if two vectors have gradients whose product is -1, they are perpendicular. The proof is immediate.

9.2 Perpendicular lines. The equation $2x - 5y = 0$ can be written as $y/x = \frac{2}{5}$, which states that the gradient of $\begin{pmatrix} x \\ y \end{pmatrix}$ is $\frac{2}{5}$.

A perpendicular vector will have gradient $-5/2$, so if (x, y) is a point on the perpendicular line through O,

$$\frac{x}{y} = \frac{-5}{2}, \quad \text{or} \quad 5x + 2y = 0.$$

An alternative way of getting this equation is interesting.

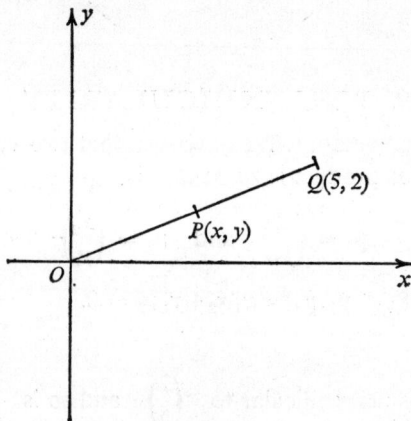

Fig. 35

If in $2x - 5y = 0$ we put $x = 5t$, then $y = 2t$. This can be written

$$\begin{pmatrix} x \\ y \end{pmatrix} = t\begin{pmatrix} 5 \\ 2 \end{pmatrix}, \quad \text{or} \quad \mathbf{OP} = t.\mathbf{OQ}$$

where Q is the point $(5, 2)$.

If (x, y) is a point of the line through O perpendicular to OQ, the scalar product of

$$\begin{pmatrix} x \\ y \end{pmatrix} \quad \text{and} \quad \begin{pmatrix} 5 \\ 2 \end{pmatrix}$$

will be zero, that is, $\qquad 5x + 2y = 0.$

308

The same argument applied to $ax+by = 0$ gives

$$\binom{x}{y} = t\binom{b}{-a}$$

for any point on the line, and, for the perpendicular line,

$$bx-ay = 0.$$

Since the lines $ax+by = 0$ and $ax+by = c$ are parallel, we have the result that, for all values of c and d, the lines

$$\left.\begin{array}{r}ax+by = c\\bx-ay = d\end{array}\right\}\quad \text{are perpendicular.}$$

Example 4. What is the equation of the line through $(2, -1)$ parallel to $x+3y = 5$?

The line has an equation of the form $x+3y = c$. Since $(2, -1)$ is a point of the line, $2+3(-1) = c$, or $c = -1$. Hence this equation is $x+3y = -1$.

This working can be abbreviated to the following:

'the parallel line is $x+3y = 2+3(-1) = -1$.'

Example 5. What is the equation of the line through $(2, -1)$ perpendicular to $x+3y = 5$?

The line has an equation of the form $3x-y = k$. Since $(2, -1)$ is a point of the line, $3(2)-(-1) = k$, or $k = 7$. Hence this equation is $3x-y = 7$.

In abbreviated style

'the perpendicular line is $3x-y = 3(2)-(-1) = 7$.'

Example 6. How far is the origin from the line through $(-1, 1)$ with gradient $-3/4$?

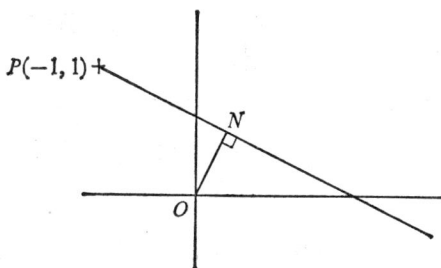

Fig. 36

The vector

$$\mathbf{n} = \binom{3}{4}$$

is perpendicular to the line and is of length 5. Hence if $\mathbf{ON} = k\mathbf{n}$, N is $(3k, 4k)$. Then

$$\mathbf{n}.\mathbf{OP} = \mathbf{n}.(\mathbf{ON}+\mathbf{NP}) = \mathbf{n}.\mathbf{ON}$$

(since **n** is perpendicular to **NP**)
$$= 5ON = 25k.$$

Hence $\quad 5ON = 3(-1)+4(1) = 1 \quad$ and $\quad ON = \frac{1}{5};$

N is the point $(\frac{3}{25}, \frac{4}{25})$.

Example 7. The circle on P_1P_2 as diameter.

What is the equation of the circle for which the points $P_1(x_1, y_1)$, $P_2(x_2, y_2)$ are at the ends of a diameter?

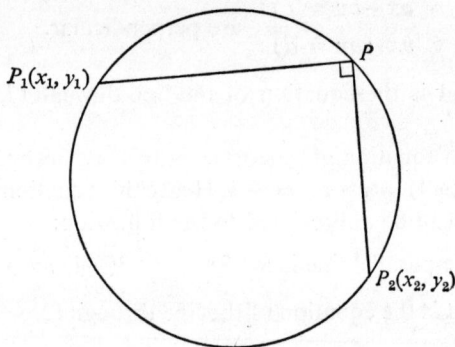

Fig. 37

If P is any point of this circle, the circle is
$$\{P: PP_1 \perp PP_2\} \quad \text{or} \quad \{P: \mathbf{P_1P.P_2P} = 0\}.$$

This is $\quad \left\{(x, y): \begin{pmatrix} x-x_1 \\ y-y_1 \end{pmatrix} . \begin{pmatrix} x-x_2 \\ y-y_2 \end{pmatrix} = 0\right\},$

that is, $\quad \{(x, y): (x-x_1)(x-x_2)+(y-y_1)(y-y_2) = 0\},$

which is the required equation.

Example 8. What is the angle between the lines
$$3x+4y = 5 \quad \text{and} \quad 8x+15y = 22?$$

The first line is perpendicular to $\begin{pmatrix} 3 \\ 4 \end{pmatrix}$ and the second to $\begin{pmatrix} 8 \\ 15 \end{pmatrix}$. The angle between them is equal to the angle between these vectors.

If
$$\mathbf{p} = \begin{pmatrix} 3 \\ 4 \end{pmatrix} \quad \text{and} \quad \mathbf{q} = \begin{pmatrix} 8 \\ 15 \end{pmatrix},$$

then $\quad p^2 = 3^2+4^2 = 25, \quad$ and $\quad q^2 = 8^2+15^2 = 289,$

so that $p = 5$ and $q = 17$.

$\mathbf{p.q} = pq \cos \theta \quad$ gives $\quad 3\times8+4\times15 = 5\times17 \cos \theta \quad$ or $\quad 84 = 85 \cos \theta$

from which $\theta = 9°$ (to the nearest degree).

310

Exercise H

1. Write down the equations of the lines through the given points parallel and perpendicular to the given lines:

(a) $(2, 3)$, $3x - y = 5$;

(b) $(-3, -1)$, $x + y = 2$;

(c) $(4, -5)$, $2x + y = 3$;

(d) $(-3, 4)$, $x = 2$;

(e) $(1, -1)$, $7x + 5y + 6 = 0$;

(f) $(0, -4)$, $13x - 8y = 17$.

2. Write down the equations of the circles with the given points at ends of a diameter:

(a) $(3, 0)$, $(0, 4)$;

(b) $(0, 0)$, $(3, 4)$;

(c) $(1, -2)$, $(-3, 5)$;

(d) $(2, -1)$, $(-3, -5)$;

(e) (p, q), $(-q, p)$;

(f) $(ct, c/t)$, $(c/t, ct)$.

3. Find the angle between the lines:

(a) $3x + y = 2$ and $4x - 3y = 1$;

(b) $2x - y = 5$ and $x + 2y + 3 = 0$;

(c) $5x + 12y + 3 = 0$ and $12x + 5y = 11$;

(d) $7x - 4y + 1 = 0$ and $y + 8x = 0$;

(e) $x - 3y - 7 = 0$ and $x + 2y + 6 = 0$.

10. TRIANGLES

Practical calculations in trigonometry depend on the existence of relations between the sides and angles of a figure, and among these relations those that refer to the sides and angles of a triangle are both the simplest and the most important. Any figure made up of line-segments can ultimately be divided into triangles; and a triangle has six elements which can be measured—the lengths of its three sides, and the magnitude of its three angles.

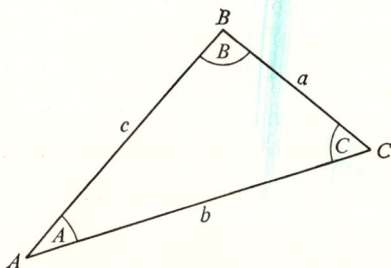

Fig. 38

We know from practical experience that we cannot choose these six elements arbitrarily; in fact, once we have chosen a suitable set of three of them, the triangle can be constructed and the other three found. There must therefore be relations which connect them.

311

The relation $\mathbf{AB}+\mathbf{BC} = \mathbf{AC}$ is of course true, but conveys no information as it stands, since it is merely the definition of vector addition. From
it, however, two very useful relations can be obtained.

We name the angles of the $\triangle ABC$ by the names of the vertices, $\hat{A}$, $\hat{B}$
and $\hat{C}$; and we use the bare capital letters A, B, and C to denote the
measures of these angles. The lengths of the sides we denote by the small
letters a, b, c:

$$a = BC \text{ (opposite } A); \quad b = CA \text{ (opposite } B);$$

$$c = AB \text{ (opposite } C).$$

10.1 The cosine rule. Since $\mathbf{BC} = \mathbf{AC}-\mathbf{AB}$, we have

$$a^2 = \mathbf{BC}^2 = \mathbf{BC}.\mathbf{BC} = (\mathbf{AC}-\mathbf{AB}).(\mathbf{AC}-\mathbf{AB})$$

$$= \mathbf{AC}.\mathbf{AC}-\mathbf{AB}.\mathbf{AC}-\mathbf{AC}.\mathbf{AB}+\mathbf{AB}.\mathbf{AB}$$

(by the distributive law for the scalar product)

$$= b^2-2bc \cos A+c^2.$$

That is, $\qquad a^2 = b^2+c^2-2bc \cos A.$

Since we might equally well have started with either of the vector
equations $\mathbf{CA} = \mathbf{BA}-\mathbf{BC}$, $\mathbf{AB} = \mathbf{CB}-\mathbf{CA}$, we must also have the
relations
$$b^2 = c^2+a^2-2ca \cos B,$$

$$c^2 = a^2+b^2-2ab \cos C,$$

which are obtained from the first by cyclic interchange of the letters:
$a \to b$, $b \to c$, $c \to a$, and similarly for the angles: $A \to B$, $B \to C$, $C \to A$.
Any general formula which is true for any triangle must still be true if
these changes are made, since it cannot depend on the way the letters are
assigned to the vertices.

These formulae enable us to calculate any side, given the lengths of the
other two sides and the angle between them.

This combination of data is usually referred to by the abbreviation *SAS*.

Example 9. Given $c = 10$ m, $a = 7$ m, and the angle $B = 37°$, calculate b.
The appropriate rule gives $b^2 = c^2+a^2-2ca \cos B$, from which

$$b^2 = 100+49-140 \cos 37°$$

$$= 149-112$$

$$= 37,$$

so that $\qquad b = 6\cdot1$ m.

Example 10. Given $A = 114°$, $b = 5\cdot8$ cm, $c = 3\cdot9$ cm, calculate a.
Here we use $a^2 = b^2+c^2-2bc \cos A$, and remember that

$$\cos (180°-\theta) = -\cos \theta.$$

Accordingly, $a^2 = 5\cdot8^2 + 3\cdot9^2 - 2 \times 5\cdot8 \times 3\cdot9 \times \cos 114°$

$$= 33\cdot6 + 15\cdot2 + 44\cdot1 \cos 66°$$

$$= 48\cdot8 + 17\cdot9$$

$$= 66\cdot7,$$

so that, $a = 8\cdot16$ cm, or $8\cdot2$ cm to 2 significant figures, the number of figures in the data.

We may also rearrange the cosine rule so as to find $\cos A$, given the three sides. From $a^2 = b^2 + c^2 - 2bc \cos A$ we derive

$$2bc \cos A = b^2 + c^2 - a^2,$$

or $$\cos A = \frac{b^2 + c^2 - a^2}{2bc}$$

and, by cyclic change, $$\cos B = \frac{c^2 + a^2 - b^2}{2ca},$$

$$\cos C = \frac{a^2 + b^2 - c^2}{2ab}.$$

Example 11. Given $a = 6$, $b = 7$, $c = 8$, find the angles.

We begin with A,

$$\cos A = \frac{7^2 + 8^2 - 6^2}{2 \times 7 \times 8} = \frac{49 + 64 - 36}{112}$$

$$= \frac{77}{112} = \frac{11}{16} = 0\cdot687, \quad \text{and} \quad A = 46\cdot6°.$$

Next, $$\cos B = \frac{8^2 + 6^2 - 7^2}{2 \times 8 \times 6} = \frac{51}{96} = 0\cdot531, \quad \text{and} \quad B = 57\cdot9°.$$

As a check, $$\cos C = \frac{6^2 + 7^2 - 8^2}{2 \times 6 \times 7} = \frac{21}{84} = 0\cdot250, \quad \text{and} \quad C = 75\cdot5°.$$

Exercise I

1. Given $a = 7$ cm, $b = 9$ cm, and $C = 49°$, find c.

2. Given $c = 83$ m, $b = 67$ m, and $A = 103°$, find a.

3. Given $a = 250$ km, $c = 88$ km, $B = 25°$, find b.

4. Calculate all the angles of a triangle, given
  (*a*) $a = 2$ cm, $b = 3$ cm, $c = 4$ cm;
  (*b*) $a = 5\cdot7$ m, $b = 8\cdot3$ m, $c = 11\cdot5$ m;
  (*c*) $a = 12\cdot5$ km, $b = 23\cdot7$ km, $c = 13\cdot8$ km.

5. Two ships are close together in mid-ocean. Ship A is proceeding at 25 knots on a course bearing 258°; ship B is proceeding due south at 21 knots. How far apart are they

  (*a*) after an hour;   (*b*) after 2 hours;   (*c*) after t hours;

assuming they maintain their courses and speeds?

6. A loop of string 14 cm long is being used to draw an ellipse. In Figure 39, A and B are fixed pins 6 cm apart; P is a pencil which keeps the loop taut and draws the curve. It can be proved that the tangent to the curve at P bisects the exterior angle APB' (see Figure 40).

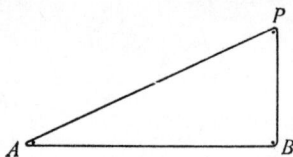
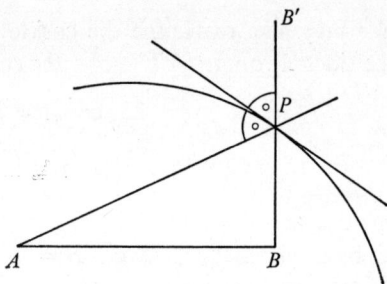

Fig. 39   Fig. 40

Calculate the angle this tangent makes with the line $\overleftrightarrow{AB}$ when $BP = 3$ cm. (First find the angles A and P in $\triangle APB$.)

10.2   The sine rule. The same vector equation $\mathbf{BC} = \mathbf{BA} - \mathbf{CA}$ can be used to give another important relation.

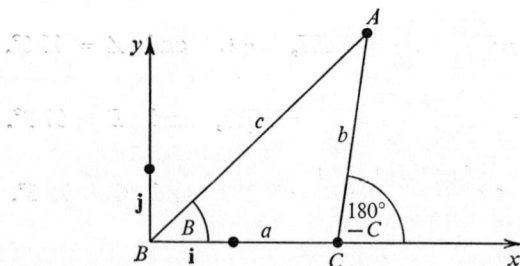

Fig. 41

If we take $\mathbf{BC}$ along the x-axis, and the y-axis perpendicular to it, then

$$\mathbf{BC} = a\mathbf{i},$$

$$\mathbf{BA} = c \cos B \, \mathbf{i} + c \sin B \, \mathbf{j},$$

$$\mathbf{CA} = b \cos (180° - C)\, \mathbf{i} + b \sin (180° - C)\, \mathbf{j}$$

$$= -b \cos C \, \mathbf{i} + b \sin C \, \mathbf{j}.$$

Hence $a\mathbf{i} = c \cos B \, \mathbf{i} + c \sin B \, \mathbf{j} + b \cos C \, \mathbf{i} - b \sin C \, \mathbf{j},$

314

and since two vectors are only the same if both their components are the same, we must have

$$a = c \cos B + b \cos C \quad \text{(component along } \mathbf{i}\text{)},$$

$$0 = c \sin B - b \sin C \quad \text{(component along } \mathbf{j}\text{)}.$$

Both these relations are useful; the second can be written

$$\frac{b}{\sin B} = \frac{c}{\sin C},$$

and, by cyclic change, each of these ratios must also equal $a/\sin A$. This is the *Sine Rule*,

$$\frac{a}{\sin A} = \frac{b}{\sin B} = \frac{c}{\sin C}.$$

Beginners often imagine that the sides of a triangle are proportional to the opposite angles; this must be wrong, for we can easily find a triangle in which $A = B + C$; for example, $A = 90°$, $B = C = 45°$. But in no triangle can we have $a = b + c$. (Why not?) The truth is given by the sine rule; the sides are proportional to the *sines* of the opposite angles. This formula enables us to find the remaining sides if we know all the angles of a triangle and one of its sides.

Example 12. Given $A = 32°$, $B = 78°$, $a = 10$ cm, find b and c.

We have
$$\frac{10}{\sin 32°} = \frac{b}{\sin 78°},$$

so that
$$b = \frac{10 \sin 78°}{\sin 32°} = 18 \cdot 5 \text{ cm}.$$

(If your slide-rule has a sine-scale (S) on it, this is very easily calculated. This scale may be on the back of the slide, or on the stock; the procedure is different in the two cases, so get someone to show you or consult the instructions issued with the rule.)

We can find C as
$$180° - A - B = 180° - 110° = 70°,$$

and then
$$c = \frac{10 \sin 70°}{\sin 32°} = 17 \cdot 7 \text{ cm}.$$

[On a Rietz pattern slide-rule these *two* calculations can be done with *one* setting of the slide (see Figure 42). First put the cursor-line to 32° on the S scale, and the 1 (that is, 10) on the C scale under the line; then, by moving the cursor in turn to 78° and 70° on the S scale, the values of b and c can be read off from the C scale.]

We can also use the sine rule to find an angle of a triangle in which we are given two sides and the angle opposite one of them.

Example 13. Given $b = 4.5$ cm, $c = 7.1$ cm, and $B = 28°$, find C.

We have

$$\frac{\sin C}{7.1} = \frac{\sin 28°}{4.5},$$

so that

$$\sin C = \frac{7.1 \sin 28°}{4.5} = 0.741.$$

Now there are two possible angles that can be in a triangle whose sines are equal to 0·741, namely 47·8° and its supplement, 132·2°. Since $c > b$, either of these will be possible and there are *two* triangles consistent with the data.

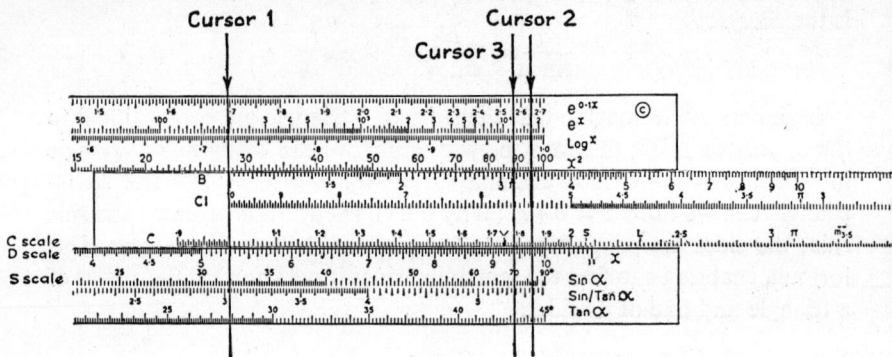

Fig. 42

We can summarize the results of this section as follows:

Given	Use
SAS	cosine rule to find a side
SSS	cosine rule to find an angle
SAA	sine rule to find two sides
ASS	sine rule to find an angle
	(? two solutions)

In every case to find a further side or angle the sine rule is used.

Exercise J

1. Why is there never any ambiguity when using the cosine rule to find an angle of a triangle?

2. Use the formulae $a = c \cos B + b \cos C$ and the two obtained from it cyclically to prove the cosine rule as follows:

$$2bc \cos A = b \cdot c \cos A + c \cdot b \cos A$$
$$= b(b - a \cos C) + c(\ldots - \ldots)$$
$$= b^2 + c^2 - a(\ldots + \ldots)$$

and deduce

$$a^2 = b^2 + c^2 - 2bc \cos A.$$

316

3. Given $A = 55°$, $B = 83°$, $b = 7 \cdot 3$ cm, find a and c.

4. Given $A = 123°$, $B = 21°$, $c = 18$ cm, find a and b.

5. Given $A = 67°$, $c = 20$ m, find possible values of B if:
   (a) $a = 25$ m;   (b) $a = 19 \cdot 2$ m;
   (c) $a = 18 \cdot 4$ m;   (d) $a = 16$ m.
Draw sketches to illustrate the various cases.

6. Calculate the answers to Exercise C, Questions 3, 5 and 6, using the methods of this section.

7. The Devil's Chair on the Stiperstones is 4·4 km from the Corndon on a bearing of 070°. Bishop's Castle is on a bearing of 167° from the Corndon and on a bearing of 205° from the Devil's Chair. Calculate how far Bishop's Castle is from the Devil's Chair. An aircraft flying on a bearing of 295° notes the bearings of Bishop's Castle, the Corndon, and the Devil's Chair as 205°, 334°, and 025°. How far is it from the Corndon and how near to it will it pass?

8. The four faces of a tetrahedron are all congruent triangles and one of its edges is 5 cm long. At the vertex at one end of this edge, the angles this edge makes with its two neighbouring edges are 38° and 75°. Calculate the lengths of all the edges of the tetrahedron.

9. A rectangular block has three edges meeting at right-angles at a corner A; $AB = 15$ cm, $AC = 36$ cm, $AD = 20$ cm. It is sawn through the vertices BCD exposing the $\triangle BCD$. Calculate the sides of this triangle, and the angle CBD.

11. THE ADDITION FORMULAE

11.1 Rotation of a vector. In Book T4, Chapter 6, it was shown that a transformation of the plane which was an isometry and kept the origin $(0, 0)$ fixed could be represented by a matrix $\begin{pmatrix} a & b \\ c & d \end{pmatrix}$, in which the first column vector was the transform of the base vector $\begin{pmatrix} 1 \\ 0 \end{pmatrix}$ and the second column vector was the transform of the base vector $\begin{pmatrix} 0 \\ 1 \end{pmatrix}$. A rotation through an angle α about the origin is of this kind; also it is clear from Figure 43 that

$$\mathbf{i} = \begin{pmatrix} 1 \\ 0 \end{pmatrix} \quad \text{transforms into the vector} \quad \begin{pmatrix} \cos \alpha \\ \sin \alpha \end{pmatrix},$$

while
$$\mathbf{j} = \begin{pmatrix} 0 \\ 1 \end{pmatrix} \quad \text{transforms into} \quad \begin{pmatrix} -\sin \alpha \\ \cos \alpha \end{pmatrix}.$$

The matrix is therefore
$$\begin{pmatrix} \cos \alpha & -\sin \alpha \\ \sin \alpha & \cos \alpha \end{pmatrix}.$$

Applying this to the general unit vector

$$\mathbf{OP} = \begin{pmatrix} \cos\theta \\ \sin\theta \end{pmatrix},$$

it is evident that rotation about O through an angle α transforms it into the unit vector

$$\mathbf{OP'} = \begin{pmatrix} \cos(\theta+\alpha) \\ \sin(\theta+\alpha) \end{pmatrix} \quad \text{(see Figure 44)}.$$

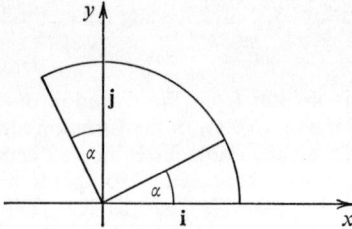

Fig. 43

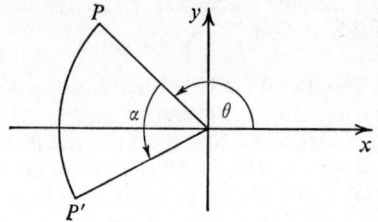

Fig. 44

Hence we have

$$\begin{pmatrix} \cos(\theta+\alpha) \\ \sin(\theta+\alpha) \end{pmatrix} = \begin{pmatrix} \cos\alpha & -\sin\alpha \\ \sin\alpha & \cos\alpha \end{pmatrix} \begin{pmatrix} \cos\theta \\ \sin\theta \end{pmatrix}$$

from which we have at once, for all angles θ and α,

$$\left. \begin{aligned} \cos(\theta+\alpha) &= \cos\alpha . \cos\theta - \sin\alpha . \sin\theta, \\ \sin(\theta+\alpha) &= \sin\alpha . \cos\theta + \cos\alpha . \sin\theta. \end{aligned} \right\} \tag{1}$$

These formulae are basic to all the relations connecting the trigonometric functions and it is difficult to overestimate their importance and usefulness. They should be second nature to anyone who hopes to acquire any skill in mathematics.

From them, writing $-\alpha$ for α, and remembering that

$$\cos(-\alpha) = \cos(360° - \alpha) = \cos\alpha$$

and

$$\sin(-\alpha) = \sin(360° - \alpha) = -\sin\alpha,$$

we have

$$\left. \begin{aligned} \cos(\theta-\alpha) &= \cos\theta\cos\alpha + \sin\theta\sin\alpha, \\ \sin(\theta-\alpha) &= \sin\theta\cos\alpha - \cos\theta\sin\alpha. \end{aligned} \right\} \tag{2}$$

The first of these we have already proved by another method in Section 8.3.

Again, if $\theta = \alpha$, equations (1) become

$$\left. \begin{aligned} \cos 2\theta &= (\cos\theta)^2 - (\sin\theta)^2, \\ \sin 2\theta &= 2\sin\theta\cos\theta. \end{aligned} \right\} \tag{3}$$

We normally write $(\cos\theta)^2$ as $\cos^2\theta$ and $(\sin\theta)^2$ as $\sin^2\theta$; hence

$$\cos 2\theta = \cos^2\theta - \sin^2\theta.$$

As in Section 8.2, or by putting $\theta = \alpha$ in the first of equations (2) we have $1 = \cos^2 \theta + \sin^2 \theta$, and hence

$$\left.\begin{array}{l} 1+\cos 2\theta = 2 \cos^2 \theta, \\ 1-\cos 2\theta = 2 \sin^2 \theta. \end{array}\right\} \tag{4}$$

11.2 Tangent formulae.

We recall that for any vector $\begin{pmatrix} x \\ y \end{pmatrix}$, $\tan \theta$ is defined as

$$\frac{y}{x} = \frac{r \sin \theta}{r \cos \theta} = \frac{\sin \theta}{\cos \theta}.$$

Hence, for all angles θ,

$$\tan \theta = \frac{\sin \theta}{\cos \theta}.$$

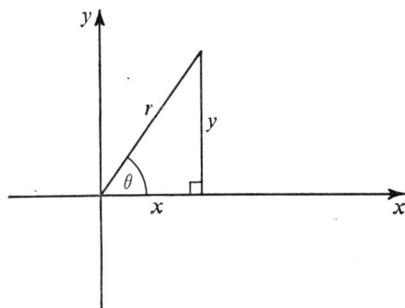

Fig. 45

From (1) we have therefore

$$\tan (\theta + \alpha) = \frac{\sin (\theta + \alpha)}{\cos (\theta + \alpha)} = \frac{\sin \theta \cos \alpha + \cos \theta \sin \alpha}{\cos \theta \cos \alpha - \sin \theta \sin \alpha}.$$

Divide numerator and denominator of this fraction by $\cos \theta \cos \alpha$, to obtain

$$\tan (\theta + \alpha) = \left(\frac{\dfrac{\sin \theta}{\cos \theta} + \dfrac{\sin \alpha}{\cos \alpha}}{1 - \dfrac{\sin \theta \sin \alpha}{\cos \theta \cos \alpha}} \right) = \frac{\tan \theta + \tan \alpha}{1 - \tan \theta \tan \alpha}. \tag{5}$$

Unlike the sine and cosine, the tangent of $(\theta + \alpha)$ can therefore be expressed entirely in terms of tangents.

Putting $\theta = \alpha$ we have

$$\tan 2\theta = \frac{2 \tan \theta}{1 - \tan^2 \theta}. \tag{6}$$

[These formulae do not hold if any of $\cos \theta$, $\cos \alpha$, or $\cos (\theta + \alpha)$ are zero, since the 'divisions' are then not legitimate; this means that none of the angles θ, α, $\theta + \alpha$ may be an odd multiple of $90°$.]

Exercise K

1. We know that sin 30° = ½ = 0·5 from Figure 46; and that

$$\cos 30° = \sqrt{3}/2 = 0·866.$$

Also $\sin 45° = \cos 45° = \sqrt{\tfrac{1}{2}} = 0·707.$

From these facts alone, compute the values of sin 15°, cos 15°, cos 75°, sin 75°, and check from your tables.

2. The tables give sin 1° = 0·017, and cos 1° is indistinguishable from 1 to three decimal places. Use these facts and those given in Question 1 to calculate

$$\sin 31°, \quad \cos 29°, \quad \sin 44°, \quad \sin 46°.$$

3. A way of computing the approximate gradient of the graph of $y = \sin x$ is to find the increment in y when x increases by 1°; this gives the approximate rate of increase in sin x per degree. Find the gradient in this way, using cos 1° = 1 and sin 1° = 0·017 to 3 decimal places, when $x = 30°$, $x = 45°$, and $x = 60°$.

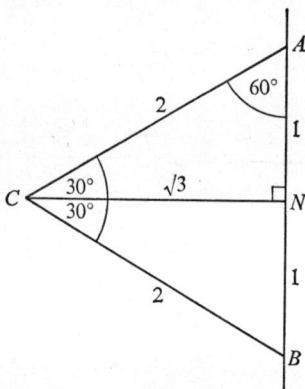

Fig. 46

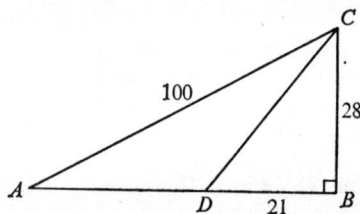

Fig. 47

4. What do the formulae (3) give for cos 120° and sin 120°, given sin 60° = $\sqrt{3}/2$ and cos 60° = ½? Give a reason why you know the answers are correct.

5. If tan $\theta \neq 0$, and tan $2\theta = -\tan \theta$, use formula (6) to get an equation for tan θ and solve it. What angles θ satisfy it in the range $0 < \theta < 360°$?

6. If sin $\theta = \tfrac{3}{5}$ and θ is acute, find cos θ, sin 2θ, and cos 2θ.

The results show that the angle in a certain Pythagorean triangle (a right-angled triangle with sides a whole number of units in length) is double the smaller angle in the 3–4–5 triangle. What are its sides?

Answer the same question starting with the smaller angle in the 5–12–13 triangle.

7. In the Figure 47, calculate the lengths of CD and AD and prove, without using any tables, that the angle at C is bisected by CD.

8. In Figure 48, if $QR = 9$ m, $SQ = 12$ m, $SP = 28$ m, calculate sin θ and cos θ as fractions without the use of tables.

9. Find simpler forms for:
   (a) $\sin(X+Y)+\sin(X-Y)$;
   (b) $\cos(X+Y)+\cos(X-Y)$;
   (c) $\cos(X-Y)-\cos(X+Y)$;
   (d) $\sin(X+Y)\cos Y-\cos(X+Y)\sin Y$.

10. Find a formula for $1+\cos 4\theta$ in terms of $\cos\theta$.

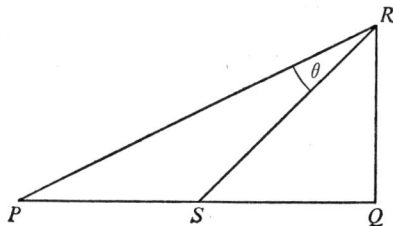

Fig. 48

11. If $\tan x = \frac{1}{5}$, calculate $\tan 2x$ and $\tan 4x$ exactly, as fractions. Is $4x$ more or less than $45°$?

12. Can you express $\sin 3A (= \sin \overline{2A+A})$ in terms of $\sin A$ only?

15

SIMPLE DYNAMICS

1. KINEMATICS

Kinematics is the study of motion. Motion is a familiar feature of everyday life, and we experience it in a great many different forms—the motion of a car on the road, of a tree in the wind, the clouds in the sky, the waves on the shore, a fly on the wall, a bird in the air, a fish in the sea, a cricket ball soaring to the boundary, a rifle bullet speeding to its target, a rocket orbiting the moon, a disc turning on the record player, couples gyrating in the dance: all is in movement, and all the movements are different. There are two essential things about motion; in the first place, it takes place in time, and in the second place it can only be observed in relation to something else. Unless we observe the apparent movement of the stars we are not aware of the turning earth. We are being carried round the sun in orbit at about 29 kilometres per second, but we never give it any thought.

For a mathematical study of motion we must therefore first consider position relative to some 'fixed' system, and then how this position changes with time.

1.1 Position. A wasp is buzzing round the classroom. How shall we fix its position? One simple way would be to say how far it is in front of, to

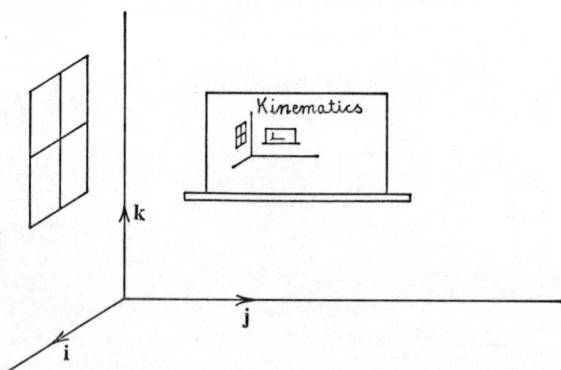

Fig. 1

the right of, and above the far left bottom corner of the room (see Figure 1). If these distances, in metres, are 5, 7, and 4, we could express the wasp's

322

position by the vector $\begin{pmatrix} 5 \\ 7 \\ 4 \end{pmatrix}$, or if we preferred it, as $5\mathbf{i} + 7\mathbf{j} + 4\mathbf{k}$, where

$\mathbf{i}, \mathbf{j}, \mathbf{k}$ are the vectors

$$\begin{pmatrix} 1 \\ 0 \\ 0 \end{pmatrix}, \begin{pmatrix} 0 \\ 1 \\ 0 \end{pmatrix}, \begin{pmatrix} 0 \\ 0 \\ 1 \end{pmatrix}:$$

the unit vectors along the three axes. Whichever way we write it, this is the *position vector* of the wasp relative to the three axes fixed in the corner of the room. The three vectors $\mathbf{i}$, $\mathbf{j}$, and $\mathbf{k}$ are called the *base vectors* in terms of which the position vector is expressed.

Example 1. The position vectors of a fly at four successive seconds are

$$\begin{pmatrix} 0 \\ 6 \\ 0 \end{pmatrix}, \begin{pmatrix} 1 \\ 4 \\ 0 \end{pmatrix}, \begin{pmatrix} 2 \\ 2 \\ 0 \end{pmatrix}, \begin{pmatrix} 4 \\ 0 \\ 0 \end{pmatrix},$$

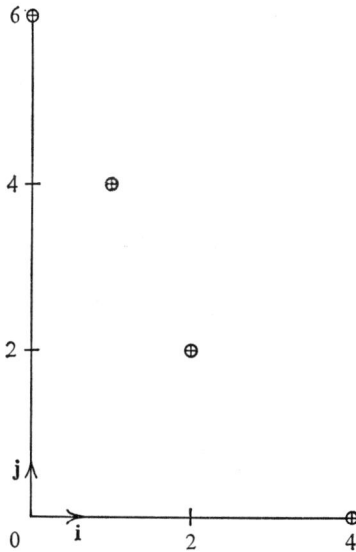

Fig. 2

for base vectors $\mathbf{i}, \mathbf{j}, \mathbf{k}$. Throughout, its distance in the direction of $\mathbf{k}$ is zero; hence the fly is moving along the floor. We can mark the four points, three of which are in a straight line, on the floor. But, without further information, we can only guess how the fly moved between these points.

323

Example 2. The position **p** of a mosquito at any time $t \leqslant 4$ is observed to be given by

$$\mathbf{p} = \begin{pmatrix} 4-t \\ 8-2t \\ 12-3t \end{pmatrix}.$$

As in Example 1, we can put $t = 0, 1, 2, 3, 4$ and plot the points obtained. But here we can also put $t = 0.1$ or any other fractional value and plot the exact course.

In fact it will be noticed that it is always twice as far to the front as it is to the right and three times as far up as it is to the right.

Its course is, therefore, a straight line towards the origin, which it reaches after 4 seconds.

Exercise A

1. A spider crawls in a straight line across the wall on the left of Figure 1. Write down the position vectors of three possible points on its path.

2. A diagram is drawn on the blackboard on the end wall. At any time t seconds after the start the position vector of the chalk-tip is

$$p = \begin{pmatrix} 0 \\ t+2 \\ \tfrac{1}{4}t^2+4 \end{pmatrix} \quad (0 \leqslant t \leqslant 3).$$

Sketch the diagram. How high must the top of the blackboard be?

3. The master stands one metre from the blackboard and drops a piece of chalk on the floor. Give the position vectors of two suitable points on its path. Could you suggest a formula for its position vector at any time t?

1.2   Velocity. Suppose we have overcome the difficulty of measuring the position of the wasp at any instant. How shall we determine its speed? If at some instant the wasp is at A (Figure 3) and a moment later it is at B, then its position has changed by the vector $\mathbf{AB} = \mathbf{b} - \mathbf{a}$ during this interval of time. We say that its

$$average\ velocity = \frac{\text{displacement } \mathbf{AB}}{\text{time taken from } A \text{ to } B}.$$

This velocity is a vector; its direction is the average direction of movement in the interval of time we are considering, and its magnitude is the average speed during this time. To find the velocity at A, however, we should have to consider shorter and shorter intervals of time, and then discover if this average velocity tended to a limit as the time-interval shrank to zero; if so, we should call this the velocity of the wasp at A. If there were a formula expressing the position vector as a function of the time, then we could obtain the velocity by differentiation. Note carefully that this velocity is

324

itself a vector; its magnitude is the speed of motion, and its direction is the direction of motion at A.

Velocity at A = rate of change of position vector at A.

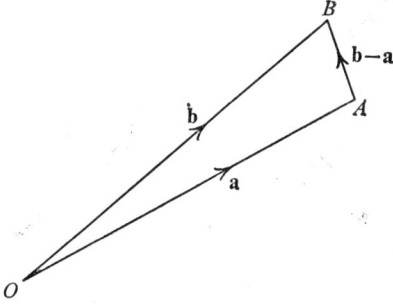

Fig. 3

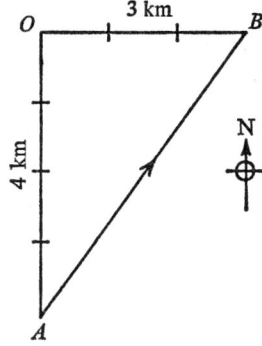

Fig. 4

Example 3. A plane flying at a negligible height is sighted 4 km away due south. One minute later it is 3 km away due east. What is its average velocity meanwhile?

By drawing or by Pythagoras, distance AB travelled = 5 km.

$$\text{Average velocity} = 5 \text{ km}/\tfrac{1}{60} \text{ hr}$$
$$= 300 \text{ km/h, in direction } \overrightarrow{AB};$$

that is, on a bearing of 36·9° (see Figure 4).

Example 4. A ship moves so that its position vector at any time t hours is

$$\binom{2t}{6t - t^2}.$$

The origin is a lighthouse and the base vectors are $\mathbf{e}$, $\mathbf{n}$ east and north, the units being in kilometres. Find its average velocity over the first 4 hour intervals, and its actual velocity when $t = 0, 1, 2, 3, 4$.

The position vectors at the end of each hour are:

$$\mathbf{OA} = \binom{2}{5}, \quad \mathbf{OB} = \binom{4}{8},$$
$$\mathbf{OC} = \binom{6}{9}, \quad \mathbf{OD} = \binom{8}{8}.$$

Displacement vectors in successive hours are

$$\mathbf{OA} = \binom{2}{5}, \quad \mathbf{AB} = \binom{2}{3}, \quad \mathbf{BC} = \binom{2}{1}, \quad \mathbf{CD} = \binom{2}{-1}$$

which are the required average velocity vectors, since the time for each

325

displacement is 1 hour. The corresponding magnitudes are $\sqrt{(2^2+5^2)}$ $\sqrt{(2^2+3^2)}$, $\sqrt{(2^2+1^2)}$, $\sqrt{(2^2+(-1)^2)}$ that is,

$$\sqrt{29}, \ \sqrt{13}, \ \sqrt{5}, \ \sqrt{5} \text{ km/h}.$$

The actual velocities are obtained by differentiating the position vector with respect to t: i.e.

$$\mathbf{p} = \begin{pmatrix} 2t \\ 6t - t^2 \end{pmatrix} \Rightarrow \mathbf{v} = \frac{d\mathbf{p}}{dt} = \begin{pmatrix} 2 \\ 6 - 2t \end{pmatrix}.$$

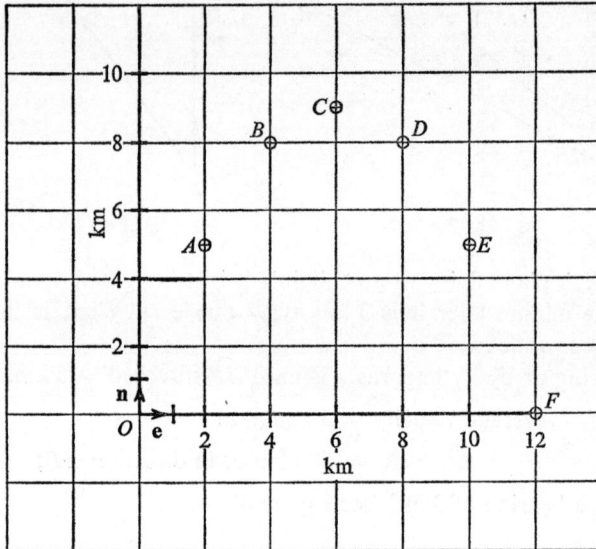

Fig. 5

From this we find that the velocities at $t = 0, 1, 2, 3$, and 4 are

$$\begin{pmatrix} 2 \\ 6 \end{pmatrix}, \ \begin{pmatrix} 2 \\ 4 \end{pmatrix}, \ \begin{pmatrix} 2 \\ 2 \end{pmatrix}, \ \begin{pmatrix} 2 \\ 0 \end{pmatrix}, \ \begin{pmatrix} 2 \\ -2 \end{pmatrix}.$$

The eastward component of the velocity is constant at 2 km/h.

Exercise B

1. A 4 m ladder leans against a wall with its foot 1 m from the bottom of the wall. It starts to slip, its foot moving $\frac{2}{3}$ m horizontally away from the wall in each second. By drawing to scale, find the average velocity of the centre of the ladder over the first and second seconds.

2. A ball moves so that its position vector at any time t seconds after starting is given by $\mathbf{p} = 32t\mathbf{i} + (64t + 16t^2)\mathbf{j}$, the axes being horizontal and vertically upwards.
    (a) Plot its path.
    (b) Find its average velocity over the first four intervals of 1 second.
    (c) Find an expression for its velocity vector, and evaluate it for $t = \frac{1}{2}, 1\frac{1}{2}, 2\frac{1}{2}, 3\frac{1}{2}$.

326

3. A particle moves so that its position vector at time t is $\begin{pmatrix} t^2 \\ 2t \end{pmatrix}$. Find its average velocity over the first three intervals of 2 seconds. Find also its velocity when $t = 0, 1, 2$ and 3.

4. The position of a missile is given by the vector $\mathbf{p} = 3\mathbf{i}+4t\mathbf{j}+5t\mathbf{k}$, where t seconds is the time after launching. Find its velocity after 2 seconds.

*5. A particle moves so that after t seconds its position vector is $\begin{pmatrix} 3 \cos t \\ 3 \sin t \end{pmatrix}$. Trace its path, and find its velocity at any time. Show that its speed is always 3 units.

1.3   Acceleration. Let us first summarize the situation so far. If at time t_A the position vector of a particle is $\mathbf{p}_A$, and at a later time t_B the position vector is $\mathbf{p}_B$, then the average velocity is the vector

$$\frac{\mathbf{p}_B - \mathbf{p}_A}{t_B - t_A},$$

and the velocity at the point A is the limit of this vector as $B \to A$. We may write this velocity

$$\mathbf{v}_A = \frac{d\mathbf{p}}{dt},$$

evaluated at t_A, using the notation of Chapter 9. We now consider the vector change in the velocity as we go from A to B. This is $\mathbf{v}_B - \mathbf{v}_A$, and it can be found from a diagram such as Figure 6, in which the velocities $\mathbf{v}_A, \mathbf{v}_B$ are represented by line-segments on a suitable scale. It should be

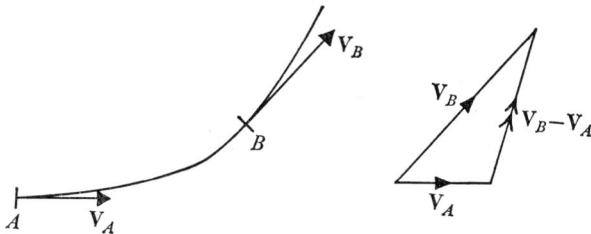

Fig. 6

clear from the diagram that the vector change in velocity is quite different from the mere change in speed; in fact, the two vectors $\mathbf{v}_A$ and $\mathbf{v}_B$ could be equal in length, so that the speeds at A and B were the same, but still have a vector difference. We now define the average acceleration from A to B as

$$\text{average acceleration} = \frac{\mathbf{v}_B - \mathbf{v}_A}{t_B - t_A}.$$

Notice that this is an extension of the idea of acceleration; it does not mean necessarily that the motion is getting faster or slower, it just means that the velocity is changing, and it can change in direction as well as in size. Again the acceleration at A is defined as the limit of this as $B \to A$; in the language of differentiation it is

$$\mathbf{a}_A = \frac{d\mathbf{v}}{dt}, \quad \text{evaluated at } t_A.$$

Example 5. Find the acceleration at any time in Example 4.

The velocity at time t is the vector

$$\begin{pmatrix} 2 \\ 6-2t \end{pmatrix}.$$

Differentiating this, the acceleration is $\begin{pmatrix} 0 \\ -2 \end{pmatrix}$. The ship has a constant acceleration of 2 km/h² to the southward.

Exercise C

Find the accelerations at any time t for the motions of Exercise B, Questions 2–5.

2. UNIFORM ACCELERATION

We shall now restrict our attention to a simple but important case of motion—that in which the acceleration is a constant vector. It can be shown that in this case the motion must all take place in one plane, which is an additional simplification, so that all the vectors concerned need only have two components.

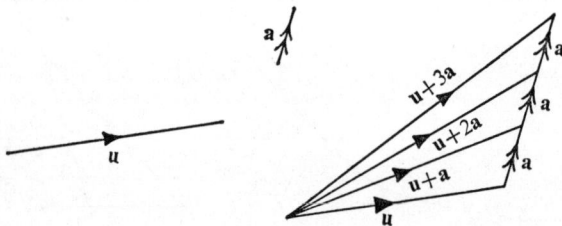

Fig. 7

If an object starts with a velocity $\mathbf{u}$ and accelerates steadily with acceleration $\mathbf{a}$, then the velocity will change by an amount $\mathbf{a}$ every second. Thus after one second the velocity will be $\mathbf{u}+\mathbf{a}$. Similarly after 2 s and 3 s the velocity will be $\mathbf{u}+2\mathbf{a}$ and $\mathbf{u}+3\mathbf{a}$ respectively as is shown in Figure 7.

In general, after a time t the velocity will be given by

$$\mathbf{v} = \mathbf{u}+t\mathbf{a}.$$

For the rest of this chapter $\mathbf{u}$ will be used to denote the constant velocity at the start of the motion under consideration and $\mathbf{v}$ will be used to denote the variable velocity t seconds later.

Furthermore, in diagrams velocities will be denoted by solid arrows and accelerations by double arrows.

Example 6. A spacecraft travelling north-east at 600 m/s starts to accelerate northwards at 10 m/s². What is its velocity after half a minute?

We have $\mathbf{v} = \mathbf{u} + t\mathbf{a}$,

$\mathbf{u}$ = 600 m/s NE,

$t\mathbf{a}$ = 300 m/s due north, since t = 30 s.

From scale drawing (Figure 8) $\mathbf{v}$ is 840 m/s in a direction 030·3°.

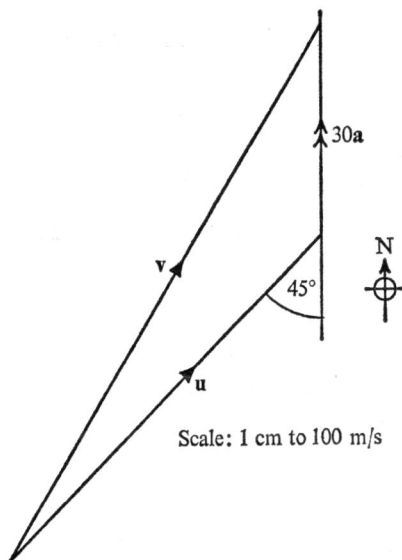

Fig. 8

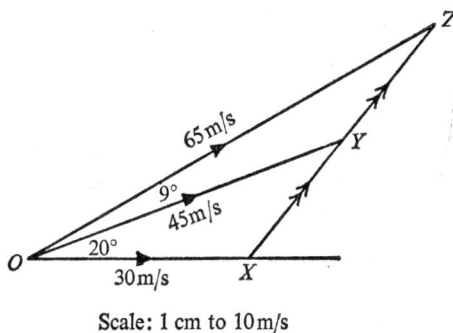

Scale: 1 cm to 10 m/s

Fig. 9

Example 7. A particle is moving in a plane with velocity 30 m/s. 4 s later its velocity is 45 m/s at an angle of 20° to its previous direction. After a further 4 s its velocity is 65 m/s at 29° to its original direction. Is this consistent with uniform acceleration?

The velocities are represented by **OX**, **OY**, **OZ** in Figure 9. Then the changes in velocity are **XY** and **YZ**. These are both found to be of magnitude 20 m/s in a direction at 50° to the original direction. Since each change took place in 4 s, this is consistent with uniform acceleration of 5 m/s² at 50° to the original direction.

2.1 Displacement with uniform acceleration. If a particle starts to move with velocity **u** and uniform acceleration **a**, then after time t its velocity is given by $\mathbf{v} = \mathbf{u} + t\mathbf{a}$.   (1)

We could have obtained this equation by integration from $d\mathbf{v}/dt = \mathbf{a}$, exactly as in Chapter 10. In exactly the same way we may obtain the displacement $\mathbf{r}$ at time t by integrating the equation giving the velocity

$$\frac{d\mathbf{p}}{dt} = \mathbf{v} = \mathbf{u} + t\mathbf{a}.$$

Put in another way, there is a function $f: t \to \mathbf{p}$ (since there is a unique vector $\mathbf{p}$ describing the position of the particle at time t), and the derived function is $f': t \to \mathbf{v}$. Therefore we conclude that

$$f: t \to t\mathbf{u} + \tfrac{1}{2}t^2\mathbf{a} + \mathbf{c},$$

that is $\qquad \mathbf{p} = t\mathbf{u} + \tfrac{1}{2}t^2\mathbf{a} + \mathbf{c}.$

At the start, $t = 0$ and $\mathbf{p} = \mathbf{c}$. Hence the displacement from $t = 0$ to t is

$$\mathbf{r} = \mathbf{p} - \mathbf{c} = t\mathbf{u} + \tfrac{1}{2}t^2\mathbf{a}. \qquad (2)$$

From this we have $\qquad 2\mathbf{r} = 2t\mathbf{u} + t^2\mathbf{a}$

$$= t(\mathbf{u} + \mathbf{v}),$$

so that $\qquad \mathbf{r} = \tfrac{1}{2}t\,(\mathbf{u} + \mathbf{v}). \qquad (3)$

Example 8. What is the displacement of the spacecraft of Example 6 in this half minute?

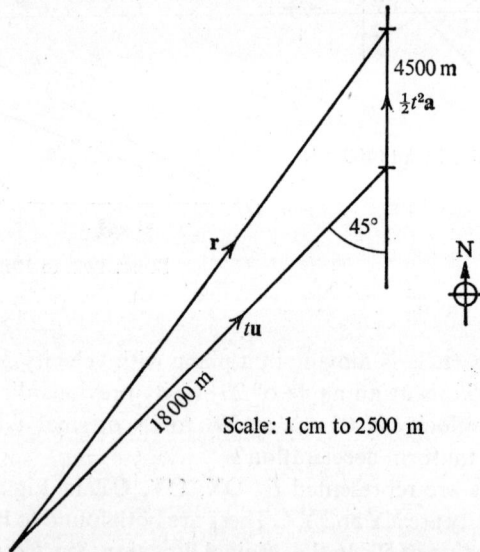

Fig. 10

Here $t = 30$ s, $\mathbf{u} = 600$ m/s north-east, $\mathbf{a} = 10$ m/s²; therefore in Figure 10,

$$\mathbf{r} = t\mathbf{u} + \tfrac{1}{2}t^2\mathbf{a}$$
$$= 18\,000 \text{ m NE} + 4500 \text{ m N}$$
$$= 21\,400 \text{ m},$$

on a bearing 036·5°.

330

Exercise D

1. A ball is thrown up at 20 m/s at an angle of 70° to the horizontal and experiences a constant acceleration of 9·8 m/s² vertically downwards. Find its velocity and distance away after 2 seconds.

2. A boat is travelling at 4 knots due east when it starts to accelerate due south at 1 knot per hour. Where is it after 3 hours? (1 knot = 1 nautical mile per hour.)

3. A rocket is observed to have a velocity of 200 m/s vertically upwards. 5 seconds later its velocity is 140 m/s at 45° to the horizontal. A further 5 seconds later its velocity is 200 m/s horizontally. Show that this is approximately consistent with uniform acceleration and find its magnitude.

4. A stone is thrown horizontally from the top of a building with velocity 20 m/s. 2 seconds later it is travelling at 45° to the horizontal with velocity 28 m/s. What is its position (i) after 2 s, and (ii) after 4 s, assuming the acceleration to be uniform?

5. An atomic particle is moving with velocity vector $\begin{pmatrix} 50 \\ 70 \end{pmatrix}$ m/s, when it experiences a constant acceleration $\begin{pmatrix} 4 \\ 2 \end{pmatrix}$ m/s². What is its velocity and change in position after 5 s?

6. A fly has a speed of 1 m/s in the direction of the vector $\begin{pmatrix} 1 \\ 2 \end{pmatrix}$ when it is accelerated with constant acceleration $\frac{1}{2}$ m/s² in the direction of the vector $\begin{pmatrix} -4 \\ 1 \end{pmatrix}$. What is its velocity after 3 s, and how far is it from its starting point?

7. Show that, if a particle is uniformly accelerated, its motion must be in a plane.

2.2   Use of the scalar product. We can establish a further useful result for uniform acceleration by the use of the scalar product defined in Chapter 14, Section 8. We begin by rewriting equations (1) and (3) of the last section:

$$\mathbf{v} - \mathbf{u} = t\mathbf{a} \qquad \text{from (1)}$$

and

$$\mathbf{v} + \mathbf{u} = 2\mathbf{r}/t \qquad \text{from (3).}$$

The scalar product of the two vectors on the left-hand side must equal the scalar product of the two vectors on the right-hand side. Accordingly

$$(\mathbf{v} - \mathbf{u}).(\mathbf{v} + \mathbf{u}) = 2\mathbf{a}.\mathbf{r},$$

or

$$v^2 - u^2 = 2\mathbf{a}.\mathbf{r}, \tag{4}$$

where u and v are the magnitudes of $\mathbf{u}$ and $\mathbf{v}$, i.e. the *speeds* at times 0 and t. The right-hand side of this equation is the magnitude a of the acceleration multiplied by the component of $\mathbf{r}$ in the direction of $\mathbf{a}$, which is a fixed direction.

Example 9. A stone is thrown with speed 10 m/s so as to hit a window 3 m up in a building 4 m away. How fast is it travelling when it hits, if it experiences a constant acceleration of 9·8 m/s² vertically downwards?

The component of **r** in the direction of **a** is -3 as the acceleration is downwards and the window is 3 m up.

Thus
$$\mathbf{a}.\mathbf{r} = \mathbf{r}.\mathbf{a} = -3\times9{\cdot}8, \quad \text{since } a = 9{\cdot}8.$$
$$\mathbf{a}.\mathbf{r} = \tfrac{1}{2}(v^2-u^2),$$
$$\Rightarrow -6\times9{\cdot}8 = v^2-10^2,$$
$$v^2 = 100-58{\cdot}8,$$
$$v = 6{\cdot}4,$$

and the speed is 6·4 m/s.

Notice that we do not need to know how far away the building is, nor the direction in which the stone was thrown so long as it hits the window.

Example 10. A bullet is fired horizontally at 300 m/s at the centre of a target 100 m away. How far below the centre does it strike if it has an acceleration 9·8 m/s² vertically downwards?

If we take axes horizontally and vertically upwards, then
$$\mathbf{u} = \begin{pmatrix}300\\0\end{pmatrix} \quad \text{and} \quad \mathbf{a} = \begin{pmatrix}0\\-9{\cdot}8\end{pmatrix}.$$

Hence if the bullet strikes x metres below the centre
$$\mathbf{r} = t\mathbf{u}+\tfrac{1}{2}t^2\mathbf{a}$$

gives
$$\begin{pmatrix}100\\-x\end{pmatrix} = t\begin{pmatrix}300\\0\end{pmatrix}+\tfrac{1}{2}t^2\begin{pmatrix}0\\-9{\cdot}8\end{pmatrix}$$

so that
$$100 = 300t$$

and
$$-x = -4{\cdot}9t^2.$$

These give immediately $t = \tfrac{1}{3}$ and $x = \tfrac{49}{90}$. The bullet drops just over 54 cm.

Example 11. A boat moves so that its distance in nautical miles east (x) and north (y) of a buoy are given at any time by
$$x = t^2-4t, \quad y = t-2, \quad t \text{ in hours.}$$
Investigate its motion for the first 5 hours.

t	0	1	2	3	4	5
x	0	-3	-4	-3	0	5
y	-2	-1	0	1	2	3

Figure 11 shows a sketch of the path of the boat. The velocity and acceleration are obtained at once by differentiation:
$$v = \begin{pmatrix}2t-4\\1\end{pmatrix}, \quad a = \begin{pmatrix}2\\0\end{pmatrix}.$$

The boat has a steady northward component of velocity of 1 knot, and a steady eastward acceleration of 2 nautical miles/h².

332

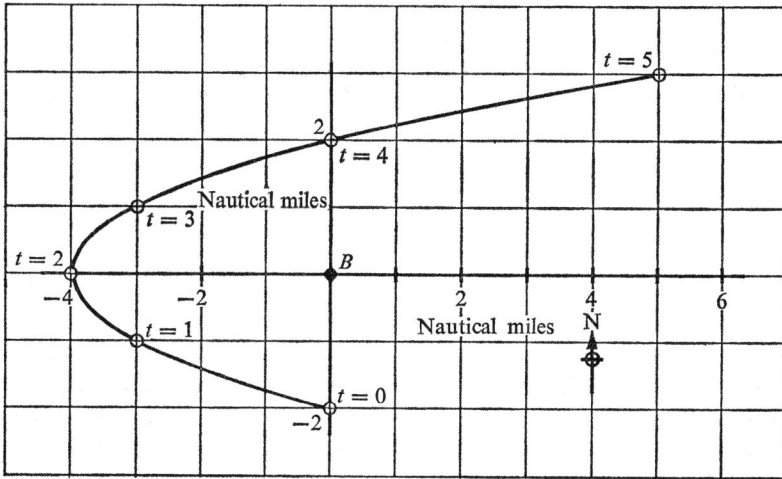

Fig. 11

Exercise E

Throughout this exercise it should be assumed that all objects experience an acceleration of 9·8 m/s² vertically downwards.

1. A ball is thrown vertically upwards at 10 m/s. How high does it rise and for how long is it in the air?

2. Water from a fire hose can just reach the top of a 30 m high building. With what velocity does it leave the nozzle?

3. How deep is a well if a stone takes 3 s to drop to the bottom?

4. A pellet is fired vertically upwards at 40 m/s. For how long will its height exceed 50 m?

5. An object takes ½ s to fall past a 3 m window. How long does it take to pass a similar window 2 m below?

6. A brick is dropped from the top of a 30 m tower at the same instant as another is thrown up from the bottom at 15 m/s. When and where will the two bricks pass?

7. A stone is thrown horizontally at 8 m/s from the top of a vertical cliff 120 m high. How far out from the base of the cliff will it land?

8. A dart is thrown horizontally at 6 m/s at the centre of a dartboard 5 m away. Where and with what velocity will it land?

9. A long jumper takes off with velocity components 4 m/s horizontally and 3 m/s vertically. How far will he jump? Would he do better if he jumped with components 4 m/s vertically and 3 m/s horizontally?

10. The position vector of an object at time t is

$$\begin{pmatrix} 24t \\ 24t - 4 \cdot 9t^2 \end{pmatrix}.$$

Investigate its motion.

3. DYNAMICS

In the previous sections we have discussed ways of describing motion, of saying *how* things move. We have not considered at all the possible causes of motion, that is to say, *why* things move. Such a study is called dynamics.

Our first attempts to give an answer to this question arise from our own immediate experience. We know that we do not move ourselves without conscious effort, that trucks do not move unless they are pushed or pulled, nor balls unless they are thrown. If a car is seen to be moving, then there must be an engine in it somehow pushing it along. We might try to make a general rule that nothing moves unless it is pushed, and everything moves when it is pushed. But we soon find exceptions to this. We let go of a cup and it falls and breaks without our pushing it; we can push as hard as we like against the wall of the house without moving it. Clouds sail across the sky without apparent effort: the wind is driving them, we say; but what drives the wind? Obviously our explanation is too simple.

It was the problem of the movement of the moon and planets which really caused this explanation to break down. Even when we remember that the apparent motion of the sky arises from the earth's rotation, there still remains the movement of the planets across the background of the stars, which is treated more fully in Chapter 18. Each planet is not attached, as was once thought, to some vast spinning crystal sphere—even if it were, what spins the sphere? Nor does it have some private angel to push it round. It was Newton (1642–1727) who first saw clearly that to look for such causes of the planets' motion is to look in the wrong place. Motion in itself, he said, does not demand a cause. Motion, once started, will persist unless something is done to stop it. What does demand a cause is *change* of motion; for example, the fact that a planet describes not a straight line, but a roughly circular orbit, so that its velocity vector is continually changing. The massive sun, Newton saw, can account for this change.

What then of our everyday experiences? When we have accelerated the car up to 40 km/h, why cannot we switch the engine off and drift to our destination? Why do we need to work to cycle a straight level kilometre? The answer lies in the need to generalize what we mean by our naïve ideas of a push; we must include a whole set of forces, which are not at first obvious, which we call *resistances*; such forces arise from contact with other material objects and with the atmosphere that surrounds us; they affect

334

only to a tiny degree the moon and planets which move in almost entirely empty space. It is only when we take such forces into account that Newton's Laws make consistent sense.

3.1 Newton's Laws. We can now state in formal language the ideas which Newton developed. He himself laid down three simple but far-reaching principles which he stated in Latin, and from which he developed the mathematical model of forces and motion which we call Newtonian dynamics.

> *Newton's First Law.* Everything continues with constant velocity unless acted on by a force.

Note that constant velocity implies constant speed (possibly zero) in a straight line. If a body does not move in this way we shall say that a force is acting on it. Thus the First Law really tells us what we are going to call a force. A force is something which causes acceleration.

> *Newton's Second Law.* A force (**P**) acting on a body is proportional to the acceleration which (**a**) it causes. In symbols

$$\mathbf{P} = m\mathbf{a}.$$

Note 1. m, the constant or proportionality, depends on the body to which the force is applied. It requires three times as much force to accelerate an object at 6 m/s² as it does to accelerate it at 2 m/s². If we applied these same two forces to a different object, the accelerations would probably be different, but one would still be three times the other. This constant m for an object is called its *mass*.

Note 2. The force **P** is wholly equivalent to $m\mathbf{a}$: it is a vector quantity in the direction of **a** and of magnitude ma.

Exercise F

1. Does it require more force to get a car moving at 20 km/h in 100 metres, or to stop it from 20 km/h in 100 metres?

2. Do you think the brakes have to exert as much force in the case of the car in Question 1 as the engine does? Explain your answer.

3. Certainly in a vacuum, and approximately in air, all dropped bodies fall to the ground with the same acceleration. What can you say about the forces acting on them?

4. If a body feels heavier to hold, is it more massive? Why?

5. The earth goes round the sun in a nearly circular orbit. Is there a force acting on the earth? In what direction do you think it is?

6. I whirl a heavy stone round on the end of a string. In what direction is the force acting on the stone?

335

7. In what direction is the force acting on a cricket ball the moment it has left the bowler's hand?

8. A squash ball bounces off a wall at the same angle as it struck it, and with no loss of speed. In what direction has a force acted on it?

3.2 Mass and force. The measurement of force is closely linked with the measurement of mass. In theory, if we have a force whose action we can repeat at will, we can apply it to a number of objects and observe their accelerations. We can then say that their masses are inversely proportional to these accelerations. If therefore we decide on one object as a standard for mass, we can determine the mass of any other object by comparison with it. A carefully preserved object has been chosen as the unit of measurement of mass; this is the standard kilogram (kg). By comparing the acceleration of other objects with that of this standard we can find the mass of any object in kilograms. The mass is a measure of the amount of matter composing the object. It is not the same as the weight of an object which may vary from place to place and which we shall discuss later. We shall then see that in fact there are simpler practical methods of comparing masses.

Forces can now be measured by applying them to a standard mass and noting the accelerations they produce. There is one standard force that we shall use:

1 *Newton* is that force which gives a mass of
1 kg an acceleration of 1 m/s².

It may be helpful to realize that a newton is approximately the force required to lift a 100 g 'weight'.

Throughout the rest of this chapter we shall normally use the following consistent system of units; when working exercises, it is best to convert all quantities into these units at the start.

	Length	Mass	Time	Speed	Acceleration	Force
S.I.:	m	kg	s	m/s	m/s²	N = kg m/s²

Example 12. What force is required to give a 1 tonne car an acceleration of 10 m/s²?

$$m = 1 \text{ tonne} = 1000 \text{ kg}, \quad P = ma$$

$$a = 10 \text{ m/s}^2. \qquad\qquad = 1000 \text{ kg} \times 10 \text{ m/s}^2$$

$$= 10000 \text{ N}.$$

In reality, considerably more force than this would be needed to overcome resistances (air, friction) to motion.

336

Example 13. A bird of mass $\frac{1}{2}$ kg is gliding due east at 4 m/s when it is blown by a wind with a constant force. By the end of 5 s its velocity has become 5 m/s in a direction E 37° S. What is the force of the wind?

From the diagram, the change in velocity **XY** = 3·0 m/s due south.

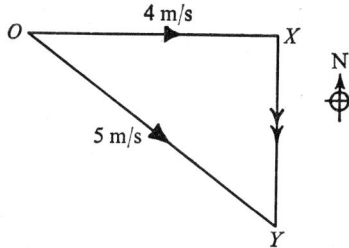

Scale: 1 cm to 1 m/s

Fig. 12

Hence the average acceleration = 0·6 m/s² due south.

$$\mathbf{P} = m\mathbf{a}$$
$$= \tfrac{1}{2} \text{ kg} \times 0\cdot6 \text{ m/s}^2 \text{ due south.}$$
$$= 0\cdot3 \text{ N due south.}$$

Example 14. A constant force of 8 N is applied to an object of mass 2 kg for 3 s. How far is it displaced if it was initially stationary?

$$\mathbf{P} = m\mathbf{a}$$
$$\Rightarrow 8 \text{ kg m/s}^2 = 2 \text{ kg} \times \mathbf{a},$$
$$\Rightarrow \mathbf{a} = 4 \text{ m/s}^2$$

in the direction of the applied force.

Also

$$\mathbf{r} = t\mathbf{u} + \tfrac{1}{2}t^2\mathbf{a}$$
$$= 0 + \tfrac{1}{2} \times 9 \text{ s}^2 \times 4 \text{ m/s}^2$$
$$= 18 \text{ m}$$

in the direction of the applied force.

Exercise G

1. What force is needed to give a 60 kg man an acceleration of 2 m/s²?

2. What force would give an α-particle, of mass 6×10^{-24} gm, an acceleration of 7×10^{12} cm/s²? How long would it need to be applied to give a velocity of half the velocity of light?

[Velocity of light = 3×10^8 m/s.]

3. A bullet of mass 30 gm is acted upon by a force of 90 N. What is its resulting acceleration? Will it break the sound barrier if the force acts for $\frac{1}{2}$ s?

4. A force of 12 N gives an object an acceleration of 6 m/s². What is the mass of the object?

5. Four aero-engines, each developing a thrust of 5000 N, give a plane of mass 72 000 kg an acceleration of 25 cm/s². What is the magnitude of the air resistance?

6. A satellite (of mass 150 kg) has a velocity of 100 m/s, which becomes one of 75 m/s at right-angles to its original direction of motion when a constant retarding force is applied for ten seconds. What is the magnitude and direction of this force?

7. An object of mass 1 kg is travelling at 20 m/s when a constant force of 5 N is applied at 150° to the direction of motion. What will the direction of motion be after 3 s?

4. GRAVITY

We have seen that from observation of the motion of a body its acceleration can be determined and hence the force acting on it.

Thus Newton, by observing motion on the earth and also of the moon, was led to the remarkable hypothesis that all objects must be attracted to each other. This force between any two objects depends only on their mass and the distance between them. So Newton propounded his *Universal Law of Gravitation*:

$$P = G\frac{mM}{d^2},$$

where **P** is the force of attraction between any two objects of masses m and M and at distance d apart. G is Newton's Universal Gravitational Constant.

The value of G is so very small, about $6 \cdot 7 \times 10^{-11}$ in S.I. units, that you will not be attracted to your desk very strongly. However, the earth, being so massive, has quite a powerful effect.

Weight. The force with which an object is attracted by the earth is called its *weight* (on earth) or the *force due to gravity*. Thus our weight depends on where we are. It is almost constant over the earth's surface. But, if we go in a space-ship (further away) or we are on the moon (less massive), we experience less attraction and are lighter. However, our mass is invariant.

By comparing $P = m(GM/d^2)$ with $P = m\mathbf{a}$ it is seen that the force of attraction on a mass m due to a mass M distant d away is just sufficient to give m an acceleration **a**, directed towards M, the magnitude of which is

$$a = \frac{GM}{d^2}.$$

The interesting thing is that this acceleration is the same whatever the value of m may be. In particular, if M is the mass of the earth and d the

338

distance of an object on the earth's surface from the centre of the earth (that is, the earth's radius R), then every object on the surface of the earth experiences the same acceleration towards the centre of the earth, of magnitude GM/R^2. This acceleration is called the *acceleration due to gravity* and is denoted by g, where

$$g = \frac{GM}{R^2}.$$

Experimental measurement has shown that, approximately:

$$g = 9 \cdot 8 \text{ m/s}^2.$$

Thus Newton was able to explain the observation made by Galileo at Pisa in the previous century.

Example 15. A man has mass 70 kg. What is his weight (on earth)?

$$P = m \times \frac{GM}{R^2}$$
$$= m \times g$$
$$= 70 \text{ kg} \times 9 \cdot 8 \text{ m/s}^2;$$

$$\text{weight} = 686 \text{ Newtons.}$$

Example 16. Calculate the mass of the earth. Take the radius of the earth to be $6 \cdot 4 \times 10^6$ metres.

$$\frac{GM}{R^2} = g$$
$$\Rightarrow M = \frac{gR^2}{G}$$
$$= \frac{9 \cdot 8 \times 6 \cdot 4^2 \times 10^{12}}{6 \cdot 7 \times 10^{-11}} \text{ kg}$$
$$\simeq 6 \times 10^{24} \text{ kg.}$$

Example 17. A ball is thrown at 20 m/s up at 40° to the horizontal. Where would it hit a vertical wall which was 22 m away? (Consider the earth's attraction only.)

It experiences an acceleration, due to gravity, of g ($= 9 \cdot 8$ m/s^2) vertically downwards.

Using $\mathbf{r} = t\mathbf{u} + \frac{1}{2}t^2\mathbf{a}$ (Section 2.1), we have

$$\mathbf{u} = \begin{pmatrix} 20 \cos 40° \\ 20 \sin 40° \end{pmatrix} \text{ m/s}, \quad \mathbf{a} = \begin{pmatrix} 0 \\ -9 \cdot 8 \end{pmatrix} \text{ m/s}^2, \quad \mathbf{r} = \begin{pmatrix} 22 \\ h \end{pmatrix} \text{ m},$$

if h m is the height above the level of the thrower at which it hits the wall. Hence

$$\begin{pmatrix} 20t \cos 40° \\ 20t \sin 40° - 4 \cdot 9t^2 \end{pmatrix} = \begin{pmatrix} 22 \\ h \end{pmatrix},$$

giving

$$t = \frac{1 \cdot 1}{\cos 40°} \quad \text{and} \quad h = 22 \tan 40° - 4 \cdot 9 (1 \cdot 1 / \cos 40°)^2$$
$$= 8 \text{ approx.}$$

The ball therefore strikes the wall about 8 m higher up.

Exercise H

Throughout this exercise assume gravitational forces to be the only forces acting.

1. Calculate the force of attraction between two 50 kg men 5 metres apart.

2. Find the weight of a 60 kg man (*a*) on earth, (*b*) on the moon.
[Take $G = 6 \cdot 7 \times 10^{-11}$ S.I. units, and for the moon assume mass $= 7 \times 10^{22}$ kg, radius $= 1 \cdot 6 \times 10^6$ m.]

3. Two objects of different mass are at the same height. Explain why:
(*a*) they will land at the same time if dropped;
(*b*) they will land at different times if thrown with the same downward force.

4. An object is dropped from the top of the Eiffel Tower, 300 m high. How fast will it be travelling when it hits the ground?

5. A stone is thrown up at 75° to the horizontal with velocity 20 m/s. In what direction is it travelling after 2 s?

6. A ball is thrown horizontally with a velocity of 15 m/s from the roof of a building 12 m high. How far away from the building does it land?

7. An object is thrown up at 25 m/s at an angle of 65° to the ground. How far away does it land?

8. A stone is thrown up at 20 m/s at 60° to the ground. It is hoped to hit a window about 10 m up a building which is about 10 m away. What is the outcome?

5. MORE COMPLICATED SYSTEMS

So far we have considered single forces acting on single bodies; furthermore, the bodies have been taken as small enough for the relative motion of their different parts to be ignored. There are three immediate steps that can be taken towards more realistic models of the situations that occur in real life:

(1) we may consider several forces acting on one body;
(2) we may consider several bodies interacting on one another;
(3) we may consider extended bodies with forces acting in different places.

We shall follow up (1) and (2), but (3) will not be considered further here.

5.1 Several forces acting on one body. There is no new problem here. By Newton's Second Law, force is a vector quantity, and all we have to do is to combine the various forces into a single force by the vector

340

addition law and the methods of Chapter 14. We then apply Newton's Second Law to this single resultant force. Alternatively, we may combine vectorially the accelerations produced by the individual forces; the result will be the same.

In the exercise that follows, scale drawing is usually an acceptable way of finding the sum of a set of vectors.

Exercise I

1. A boy and a girl pull a 5 kg toboggan in directions which differ by 40°; the boy pulls with force 18 N and the girl with force 15 N. What acceleration would they give the toboggan if there were no resistance to motion?

2. A ship is towed by three tugs; one pulls due north with a force of 200000 N, one pulls north-west with 300000 N. The third pulls north-east with 400000 N. In what direction does the ship move?

3. Two boys try to pull a 25 g ball-bearing by means of magnets; one pulls with a force of 0·1 N, the other pulls at right-angles to this with a force of 0·15 N. What is the acceleration if there is a resistance to motion of 0·1 N?

4. A 1000 kg helicopter takes off with vertical lift of 15000 N for 5 seconds. How far does it rise? (Take $g = 10$ m/s².) If it then moves horizontally with a thrust whose horizontal component is 12000 N for 5 s, how far does it move horizontally and how far is it then from where it took off? By calculating the resulting effect, find how far away it would have been after 5 s if it had taken off using horizontal and vertical thrust simultaneously.

5. A satellite of mass 50 kg is attracted by the earth with a force of 250 N and, at an angle of 120° to this, the sun attracts it with a force of 50 N. What is its resulting acceleration (*a*) towards the earth, (*b*) towards the sun?

6. A 100 g rocket gives a thrust of 2 N and is set off at 10° to the vertical. What is its height after 3 s? (Take $g = 9·8$ m/s² and assume no resistance.)

5.2   Several bodies. If there are two bodies not in contact, which are not electrically charged, the only force between them will be the mutual gravitational attraction GMm/d^2 discussed in Section 4. This force acts on both bodies: body *A* exerts it on body *B*, and body *B* exerts an equal and opposite force on body *A*. *A* attracts *B* towards *A*, and *B* attracts *A* towards *B*. But this force is exceedingly small unless the bodies are very massive, and we can safely neglect it as far as events on the earth's surface are concerned, except of course for the attraction of the earth itself which we call weight.

But when one body is in contact with another, further forces are called into play. There is a force exerted when a book rests on a table. The table pushes upwards on the book, and the book presses downwards on the table. Again, these forces are equal and opposite. If the table is level the forces will be at right-angles to the table—i.e. vertical—and the 'double-ended' force is called the *normal contact force* ('normal' here means 'at right-angles'). If the book is pushed so as to slide over the table there will be a force resisting this motion; it acts backwards on the book, and the book exerts a forward force on the table. This is the *frictional* contact force.

Again, when bodies are connected by a coupling or chain, the normal contact force will appear as a pull between the bodies (see Figure 13); we then call it a *tension*. If there is an intermediate string or rope, then each body in reality is exerting a force on the rope and the rope is exerting a

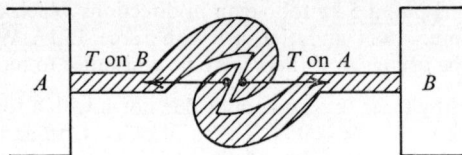

Fig. 13

force on the bodies. But because the rope is usually much less massive then the bodies, the forces on it add up to zero ($P = 0$ if $m = 0$, whatever **a** may be) or nearly so, and once again we can suppose that the bodies exert equal and opposite forces on one another. These facts are summed up in

> *Newton's Third Law.* Whenever a body A exerts a force on a body B, then body B exerts an equal and opposite force on body A.

Note 1. This statement is independent of any motion of A and B, or motion relative to one another; it is true in all circumstances.

Note 2. The statement is true whether the forces are contact forces, or actions at a distance, which may be gravitational forces, electrostatic forces, magnetic forces, or the quantum-mechanical forces which hold atomic particles together; *all* forces are double-ended.

Note 3. If the force exerted by A on B is called the *action*, then the equal and opposite force exerted by B on A is called the *reaction*. The words are in common use, but which is assigned to which force depends on the point of view, and an additional objection to their use is that action has another technical meaning.

5.3 Additivity of mass. Suppose a force P is applied to two bodies in contact, of masses m_1 and m_2 (Figure 14). There will be a contact force between them, of magnitude R, say. If they remain in contact they will share a common acceleration a. For m_1, Newton's Second Law gives

$$P - R = m_1 a,$$

and for m_2, $$R = m_2 a.$$

342

Adding these equations, we obtain

$$P = (m_1 + m_2)\, a.$$

In other words, the two bodies behave like a single body of mass $m_1 + m_2$. In general, we can say that if a body is broken up into parts, its mass will be the sum of the masses of its parts. This is why we often say, loosely, that mass is the quantity of matter in a body.

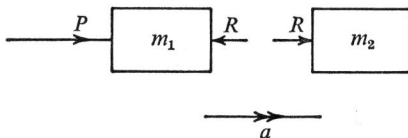

Fig. 14

We use this principle whenever we use an ordinary balance to determine mass. We assume without question that the mass of a 50 gm 'weight' and a 10 gm 'weight' placed together on the scale pan is 60 gm. That this is so is a consequence of Newton's Third Law.

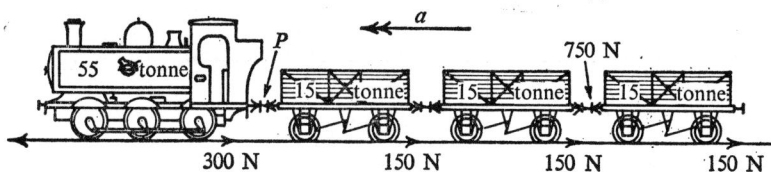

Fig. 15

Example 18 (see Figure 15). A locomotive is pulling three trucks of mass 15 tonnes each. The tension in the last coupling is 750 N. If the mass of the locomotive is 55 tonnes and the resistance to motion is 150 N for each truck and 300 N for the loco., find the acceleration and the tension in the first coupling.

Consider the last truck. The forces acting on it (horizontally) are 750 N tension in the coupling and 150 N resistance. If the acceleration is a, then

$$(750 - 150)\,\text{N} = 15 \text{ tonnes} \times a$$

$$\Rightarrow 600 \text{ N} = 15000 \text{ kg} \times a$$

$$\Rightarrow a = \frac{600}{15000}\,\frac{\text{N}}{\text{kg}}$$

$$= \tfrac{1}{25}\,\text{m/s}^2, \quad \text{since } 1\,\text{N} = 1\,\text{kg m/s}^2.$$

343

Now the three trucks behave like a single body of mass 45 tonnes. Hence if P is the tension in the first coupling,

$$P - 450 \text{ N} = 45 \text{ tonnes} \times \tfrac{1}{25} \text{ m/s}^2$$

$$= \frac{45 \times 1000 \times 1}{25} \text{ kg m/s}$$

$$= 1800 \text{ N}$$

$$\Rightarrow P = 2250 \text{ N}.$$

It is obvious on common-sense grounds that the tension in the first coupling is three times that in the third.

Example 19. A 2 tonne lorry is facing down a hill sloping at 20° to the horizontal. The brakes provide a resistance of 6500 N. What will its acceleration be? (see Figure 16).

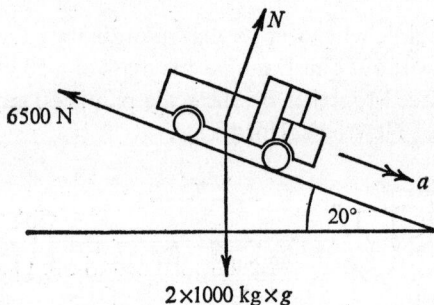

Fig. 16

There will be a normal contact force N in addition to the weight and the frictional resistance produced by the brakes. To keep N out of the equations, we consider components of forces at right-angles to N. (This is a sound general principle.) This gives

$$2 \times 1000 \times g \times \sin 20° - 6500 \text{ N} = 2 \times 1000 \text{ kg} \times a$$

$$(6700 - 6500) \text{ N} = 2000 \text{ kg} \times a$$

$$\Rightarrow a = \frac{200}{2000} \frac{\text{N}}{\text{kg}}$$

$$= \tfrac{1}{10} \text{ m/s}^2.$$

Exercise J

1. What is the equal and opposite reaction to your weight?

2. When you climb a rope, what force pulls you up? How do you bring this force into being?

3. When a bicycle is being accelerated, in what direction is the force of friction acting on the bicycle:
(*a*) between the rear wheel and the ground;
(*b*) between the front wheel and the ground?

4. Two boys pull opposite ways on a rope with a force of 150 N each. What is the tension in the rope?
If one boy attached the rope to a hook in the wall and pulled on it with the same force, what would be the tension in it now?

5. Draw a diagram to show the forces acting on an aircraft in steady level flight.

6. The driver of a van containing pigeons banged on the wall of the van before crossing a flimsy bridge in order to make the pigeons take off. Did this procedure help?

7. A rocket propels itself by ejecting material backwards. Explain where the propulsive force comes from. Would the rocket perform better or worse in a vacuum (assuming that it does not need the air to burn its fuel)?

8. What acceleration does the earth (mass about 6×10^{24} kg) experience when a 2000 kg helicopter hovers above it?

9. A 1-tonne car pulls a 2-tonne caravan along a level road. If the resistance may be taken to be 100 N/tonne for any vehicle, find the tension in the coupling:
(*a*) at a steady speed; (*b*) when accelerating at 2 m/s².

10. A plane of mass 3000 kg tows a 200 kg glider. It is observed that they are both travelling horizontally at 90 m/s with the towing cable inclined at 30° to the horizontal. It is assumed that the lift on the glider is ¾ of its weight and on the plane the whole of its weight. The horizontal drag is the same for the glider and the plane. Find
(*a*) the tension in the cable; (*b*) the resistance;
(*c*) the magnitude and direction of the thrust of the plane's engines.
(Take $g = 10$ m/s².)

11. A toy truck of mass 500 gm is pushed up a plane inclined at 30° to the horizontal by a force of 4N. Find its acceleration, neglecting resistances.

12. If the same truck runs down the plane against a resistance of 2N, what will its acceleration be?

13. The same truck is observed to run with constant speed down a slope of 16°. What are the resistances to its motion?

14. In Example 18, find the propulsive force on the locomotive. Where does it come from? What is the ratio (frictional force/normal force) between the driving wheels and the rail? (All wheels are coupled.)

6. MATHEMATICAL MODELS

Newton's three laws, in principle, enable us to solve problems of all kinds relating to motion. In practice, however, any problem is very complicated, involving many forces of different kinds, and simplifying assumptions have to be made. Producing a simple, easily soluble version of a physical situation is described as making a *mathematical model*.

Among assumptions frequently made are the following:

(1) Resistances are zero, or if not, at least constant.

(2) Bodies are concentrated masses, and the relative movement of their parts is unimportant. For example, we neglect the forces needed to turn the wheels of trucks and cars.

(3) Pulleys are smooth, and require no effort to turn them. The tension in a string passing over such a pulley is the same on both sides.

(4) Strings have no mass, and can be bent without force.

(5) Forces do not vary during the time under discussion.

These assumptions can of course be relaxed, but the resulting mathematics becomes correspondingly more complicated.

The following example illustrates the procedure.

Example 20. A lift hangs freely, balanced by a counterweight of equal mass 150 kg. A man, mass 75 kg, steps in. What would his acceleration be if there were no resistance to motion?

We assume that the counterweight, also the man in the lift, are small blobs of mass 150 kg and 225 kg.

The forces acting are in order of magnitude

  (i) the attraction of the earth,

  (ii) the pull of the cable;

  (iii) the resistance caused by rubbing against the lift shaft;

  (iv) the air resistance.

We neglect (iii) and (iv) and indicate the others on a diagram.

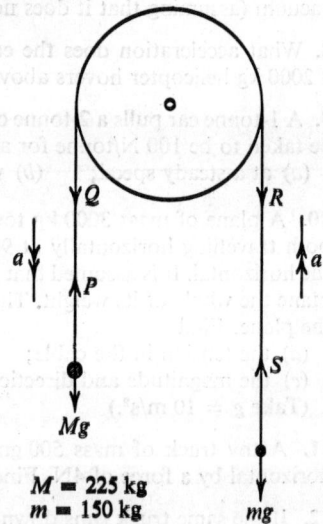

$M = 225$ kg
$m = 150$ kg
Fig. 17

P, Q are the cable's pull on the lift and counterweight.

R, S are the cable's pull on the pulley support.

We now use further assumptions:

  (v) the force to turn the pulley is negligible;

  (vi) the mass of the cable is negligible.

It requires no force to accelerate an object with no mass, so the resultant force on any portion of cable is zero.

Hence $$P = Q \text{ and } R = S;$$
also (v) implies $R = Q$.

Thus $P = Q = R = S$ and the 'tension' in the cable is the same throughout its length; an assumption that could have been made initially.

346

Thus for the lift:
$$225 \text{ kg} \times g - T = 225 \text{ kg} \times a,$$

where T is the tension in the cable; and for the weight:
$$T - 150 \text{ kg} \times g = 150 \text{ kg} \times a.$$

Hence $T = 180 \text{ kg} \times g = 1760 \text{ N}$ and $a = g/5 = 1 \cdot 96 \text{ m/s}^2$.

By neglecting or simplifying most of the factors affecting motion we get a basic model giving a clear picture.

All the factors can, in theory, be taken into account to give increasing precision to our model and detail to our picture. However, the work involved, though it may well be simple in principle, is often very laborious in practice. Thus, there is still the need for the physical scale model, the wind tunnel and the test flight; though the increasing use of computers has improved the situation considerably.

Our model, even so, needs considerable modification to deal with many astronomic phenomena (Relativity) and the motion of atomic particles (Quantum Mechanics). But artificial earth satellites are a spectacular justification for a model of the universe based on the simple, universal predictive laws of Newton.

Exercise K

1. In each of the following, list in order of importance, the factors which influence the motion. Then illustrate simplified mathematical models which give first and second approximations to the motion of:

    (*a*) a ball;   (*b*) a boat;   (*c*) a bicycle;
    (*d*) a bullet;   (*e*) a hovercraft;
    (*f*) a helicopter;   (*g*) the moon.

2. A 70 kg workman, standing on the ground, tries to lower a 78 kg bucket of rubble from a height of 12 m by means of a pulley, and is pulled off his feet. Calculate his upward acceleration and the speed with which he hits the bucket (halfway up).

3. A packing-case, of mass 100 kg, rests on the rough floor of a loft. There is a rope round it, the end of which hangs down through a trap-door. A boy weighing 60 kg hangs on the rope, thereby accelerating the packing-case. After he has dropped 2 m on the rope he lets go and jumps aside. The packing-case has 1 m more to go, and the frictional resistance to its motion throughout is 250 N. Will it drop through the hole?

4. A winch is used to raise buckets from a well. A rope is wound round the winch and carries a bucket at each end. An empty bucket weighs 1 kg, and each bucket holds 5 litres of water when full. While the winch is running the tension on one side of the rope is 60 N more than that on the other side. How long will it take to raise a full bucket from a well 30 m deep, using the empty bucket as a counterpoise?

7. MOMENTUM

If we know the force on a body we can find its acceleration and, if we know its acceleration, we can find its change in velocity.

It is sometimes useful to combine the results (assuming **P** constant)

$$\mathbf{P} = m\mathbf{a} \quad \text{and} \quad \mathbf{v} = \mathbf{u} + t\mathbf{a}$$

to show the way in which a force, acting for a time t, changes the velocity:

$$t\mathbf{P} = m\mathbf{v} - m\mathbf{u}.$$

It is convenient to be able to talk about the applications of this equation and so special names are given to the expressions in it.

The effect $t\mathbf{P}$ of a force **P** acting for time t is called its *impulse*. It is a vector quantity, measured in s N.

The quantity $m\mathbf{v}$, the product of the mass and velocity of a body, is called its *momentum*. It is a vector quantity measured in kg m/s.

Thus an impulse ($t\mathbf{P}$) causes a change in the momentum (from $m\mathbf{u}$ to $m\mathbf{v}$) and the amount by which the momentum is changed is a measure of the size of the impulse.

Example 21. A car travelling at 30 m/s runs into a brick wall. What is the average force on a 70 kg passenger if:

(a) he is brought to rest in 1/100 s;

(b) he is brought to rest in $\frac{1}{2}$ s, because he is wearing a safety belt which takes time to stretch?

The change of momentum $= m\mathbf{v} - m\mathbf{u} = 0 - 70 \text{ kg} \times 30 \text{ m/s}$

$$= -2100 \text{ kg m/s}.$$

(a) Hence $\mathbf{P} \times \frac{1}{100} \text{ s} = -2100 \text{ kg m/s}$

$$\mathbf{P} = -210000 \text{ kg m/s}^2 = -210000 \text{ N},$$

which is a force sufficient to lift about 21 tonnes.

(b) With a safety belt

$$\mathbf{P} \times \frac{1}{2} \text{ s} = -2100 \text{ kg m/s}.$$

$$\mathbf{P} = -4200 \text{ N},$$

which is a force sufficient to lift about 400 grams.

The negative sign indicates that the force acts against the direction of motion.

Example 22. In a game of French Cricket the $\frac{1}{6}$ kg ball travelling at 12 m/s is hit back with velocity 8 m/s. Find its change in momentum if its final direction is:

(a) towards the bowler;

(b) at 60° to the direction in which it was travelling.

348

Also find the average force acting in each case if the contact between bat and ball lasted for $\frac{1}{120}$ s.

(a) $t\mathbf{P} = m\mathbf{v} - m\mathbf{u}$

$\quad\quad = \frac{1}{8} \times 8 - \frac{1}{8} \times (-12)$ measuring towards the bowler.

Change in momentum $= \frac{10}{3}$ units towards the bowler. Hence $\mathbf{P} = 400$ N towards the bowler.

(b) A velocity **OX** of 12 m/s is changed to a velocity **OY** of 8 m/s. (See Figure 18.) By drawing, the change in velocity $\mathbf{XY} = 16\cdot7$ m/s, at 23° to direction of bowler. Hence the impulse $= 2\cdot8$ units at 23°, and the average force $= 336$ N.

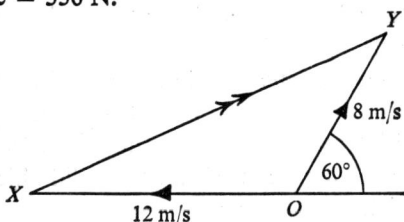

Fig. 18

Exercise L

1. A 1-tonne car and a 15 kg bicycle are both travelling at 10 m/s when they both apply their brakes so as to get a resistance of 150 N. How long will it take each to stop?

2. A 7 kg shot is put with a force of 50 N. What velocity does this give the shot if the force is applied for   (a) $\frac{1}{10}$ s;   (b) 1 s?

3. A jet of water emerges at 10 m/s from a nozzle of radius 2 cm. What mass of water emerges in 1 s if the density of water is taken to be 1 g/cm³? If this is stopped by a wall, what force does it exert on the wall? Does the direction in which it is travelling matter?

4. A bird of mass 200 g is flying at 5 m/s. $\frac{1}{4}$ s later it is flying at 7 m/s at 12° to its original direction of motion. What is its change in momentum? What is the magnitude and direction of the average force it applies to effect this change?

5. A space-ship is travelling at 200 m/s when its retro-rockets give it an impulse of 3 million N-seconds at right-angles to its direction of motion. What is its resulting velocity if its mass is 50000 kg?

6. A hockey ball travelling at 8 m/s is deflected through 35° by an impulse applied at right-angles to its direction of motion. Find its resulting velocity.

7. A beam of electrons is deflected through 20° by the deflector plate of a cathode ray tube, its speed being unaltered. Find the magnitude of the change in momentum as a percentage of the magnitude of momentum before deflection.

8. SYSTEMS

Internal forces. If you wanted to calculate how high you could jump, your main concern would be with the earth's attraction and the force of your feet on the ground. The interaction between your head and your neck, as with all the other equal and opposite interactions holding you together, would have no effect.

Any interaction inside a system which we are considering as a whole is called an *internal interaction* and has resultant zero.

Any other force acting on this system is called an *external force*, since it must be caused by interaction between the system and some other external object, such as the earth.

This is of use when we are interested in the overall motion of a system rather than that of its constituent parts. We need only consider the external forces, as the internal interactions will have no effect. In fact, this is our main justification for applying Newton's laws to solid objects and treating these objects like particles.

Collision. Consider what happens when a stationary shell explodes. All the forces are internal interactions. Thus their resultant effect is zero! However, this does not mean that nothing happens. Each force acts for a time and changes the momentum of some portion of the shell. The portions will move off in all directions, but the vector sum of all their momenta is zero. For example, there will be as much momentum backwards as there is forwards.

The same thing happens when two objects collide. Overall, the momentum is unchanged but it will probably be redistributed.

Thus, if no external forces act on a system, its total momentum is not changed:

Momentum at the end = momentum at the beginning;

or $$m_1 v_1 + m_2 v_2 = m_1 u_1 + m_2 u_2$$

for a system of two objects with masses m_1 and m_2.

This is used in the principle of rocket propulsion. By ejecting a small mass of fuel with a high velocity backwards a large rocket is given a small velocity forwards; the extra momentum of the rocket forwards being numerically equal to the momentum of the fuel backwards.

Similarly, when a bat hits a ball, the bat loses as much momentum as the ball gains; though extra momentum may be given by the external force of your arm acting during the time of contact.

Units. When the momentum is not changed, as in the expression above, all the terms are a measure of momentum; and any units for momentum may be used so long as they are the same throughout.

350

Example 23. A 200-tonne express train, travelling at 100 km/h, runs into the rear of a 40-tonne diesel-car travelling at 50 km/h. Find their common velocity, assuming they move on together.

Measure mass in tonnes and velocities in km/h.

Horizontally. No external forces act on the two-train system. Hence momentum at end = momentum at beginning; i.e.

$$(m_1+m_2)v = m_1u_1+m_2u_2,$$
$$(200+40)v = 200 \times 100+40 \times 50$$
$$= 22000,$$
$$v = 90 \text{ km/h.}$$

Example 24. An electron travelling at one-third of the velocity of light hits a stationary neutron. It is deflected through 70° and its velocity halved by the collision. Find the direction in which the neutron moves and its velocity.

(Take the mass of the electron to be 10^{-30} kg, the neutron 2×10^{-27} kg, and the velocity of light to be 3×10^8 m/s.)

Take the velocity of light as unit velocity and the mass of the electron as unit of mass.

Initial speed = $\frac{1}{3}$ unit⎫
Final speed = $\frac{1}{6}$ unit⎭ for the electron; ⎰electron mass = 1 unit,
⎱neutron mass = 2000 units.

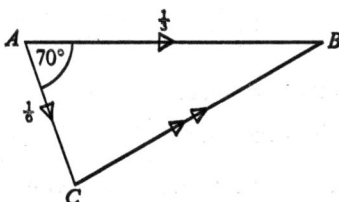

Fig. 19

In Figure 19,

AB represents initial momentum of electron: $1 \times \frac{1}{3} = \frac{1}{3}$ units.

AC represents final momentum of electron: $1 \times \frac{1}{6} = \frac{1}{6}$ units.

Since there is no change in total momentum, **CB** represents the final momentum of the neutron.

Its direction is at 29° to line of approach of electron, and its magnitude

$$= 0 \cdot 32 \text{ units}$$

Hence the neutron's velocity

$$= \frac{0 \cdot 32}{2000} \times \text{velocity of light}$$

$$= 4 \cdot 8 \times 10^4 \text{ m/s.}$$

This model is very inaccurate because we have neglected relativistic and other effects.

We end with a thought for the space-age.

If your car breaks down you can push it along; but what happens if your space-ship breaks down and you start to push it?

You cause interaction, so that the force of the ship on you is equal and opposite to your force on the ship (Newton's Third Law).

These forces cause you to accelerate apart (Newton's Second Law).

Finally, you will lose contact, interaction will cease, and you will continue to drift apart with uniform velocity (Newton's First Law).

Happily this is not quite true as your small mutual attraction will ultimately bring you together again (Newton's Law of Gravitation).

Exercise M

1. A 10 g bullet is fired at 300 m/s from a 3 kg rifle. Find the velocity of recoil of the rifle. What would it be if the rifle were firmly held by a 70 kg boy?

2. A 1-tonne car travelling at 50 km/h runs into a stationary 200 kg cow. What is their common velocity after the collision?

3. The empty fuel tanks of a rocket are jettisoned at a relative velocity of 20 m/s. If they account for three-quarters of the rocket's mass, what additional velocity does this give to the remainder?

4. A 60 kg man dives, at 4 m/s, from the back of a stationary 80 kg boat. Find the horizontal velocity with which the boat moves off if he dives:
  (a) horizontally;
  (b) upwards at 20° to the horizontal.

5. A stationary nucleus disintegrates, a third of it flying off at 6×10^4 m/s. What will be the velocity of the other portion? What would it have been if the nucleus had originally been moving at 2×10^4 m/s at 45° to the direction in which the smaller portion flies off?

6. A 60 kg boy running at 3 m/s collides with a 45 kg boy running at 5 m/s. Find their common velocity immediately afterwards if they collide:
  (a) 'head-on';   (b) at right-angles.

7. A 90 g ball, travelling at 4 m/s, is hit by a 700 g stick travelling with a velocity of 6 m/s, and this brings the stick to rest. How fast will the ball then travel if it was hit:
  (a) straight back in the direction it came from;
  (b) at 75° to the direction if came from?

8. By considering components of the total momentum prove that:
  (a) when an object disintegrates into two portions they will move off in the same line;
  (b) when an object disintegrates into three portions they will move off in the same plane.

Miscellaneous Exercise

1. The position vector of a particle at time t is $\mathbf{p} = t\mathbf{i}+t^2\mathbf{j}+t^3\mathbf{k}$, where $\mathbf{i}$, $\mathbf{j}$ and $\mathbf{k}$ are unit vectors along the three coordinate axes. Show:

(a) that at $t = 0$ it is moving along the x-axis;

(b) that for large values of t it is moving almost parallel to the z-axis;

(c) that its acceleration has a constant component and one that increases linearly with it.

Find the position and velocity of the particle at the moment when these two components are equal.

2. A particle moves on a horizontal table so that at time t its position vector relative to a point O of the table is given by

$$\mathbf{p} = 3\mathbf{i} \sin \pi t/5 + 4\mathbf{j} \cos \pi t/5.$$

Find its velocity and acceleration at time t, and show that its acceleration is always towards O. What kind of force would be needed to maintain this motion, and how could it be provided?

3. A body of mass 4 kg moves in a straight line OA, starting at O when the time t is zero. It has a velocity v m/s in the direction of OA given by the following function:

$$0 \leqslant t \leqslant 5, \quad v = 2t,$$

$$5 \leqslant t \leqslant 10, \quad v = 15-t,$$

$$10 \leqslant t, \quad v = 0.$$

Sketch this function.

Express (a) the distance travelled from O, and (b) the force acting on the body, both as functions of time. How are these results related to the sketch? Comment on what happens when $t = 10$. *(O.C.)*

4. A projectile of mass 5 kg is fired with speed 800 m/s at an elevation of 30°. After how long will it be travelling at right-angles to its original direction? At this instant it strikes the ground and is brought to rest: what is the impulse on the projectile? (State your units.) *(O.C.)*

5. A train of 30 trucks, each of mass 10 tonnes, is being hauled up an incline of 1 in 100 (i.e. the sine of the angle of slope is 0·01) by a locomotive of mass 80 tonnes. There is a resistance of 100 N/t acting down the slope on every vehicle in the train. If the speed of ascent is constant, what pull is the locomotive exerting at its driving wheels? If the last 10 trucks are detached, what will be the acceleration of the remainder, the pull remaining the same?

6. A car of mass 1000 kg starts from rest and accelerates steadily at 1·2 m/s² for 20 s; for the next 10 s it travels at constant speed, and in a further 30 s it decelerates steadily to rest. Exhibit these facts on a velocity-time diagram, and find:

(a) the total distance travelled, in metres;

(b) the deceleration;

(c) the effective force acting on the car at each stage. *(O.C.)*

9. SUMMARY

Average velocity $= \dfrac{\text{change in position vector}}{\text{time taken}} = \dfrac{\mathbf{p}_2 - \mathbf{p}_1}{t_2 - t_1}.$

Velocity $=$ rate of change of position vector $= d\mathbf{p}/dt.$

Average acceleration $= \dfrac{\text{change in velocity vector}}{\text{time taken}} = \dfrac{\mathbf{v}_2 - \mathbf{v}_1}{t_2 - t_1}.$

Acceleration $=$ rate of change of velocity vector $= d\mathbf{v}/dt.$

For uniform acceleration:

$$\mathbf{v} = \mathbf{u} + t\mathbf{a},$$
$$\mathbf{r} = \tfrac{1}{2}t(\mathbf{u} + \mathbf{v}),$$
$$\mathbf{r} = t\mathbf{u} + \tfrac{1}{2}t^2\mathbf{a},$$
$$v^2 = u^2 + 2\mathbf{a}.\mathbf{r}$$

Law of Gravitation:

$$P = GMm/d^2, \quad G = 6{\cdot}67 \times 10^{-11}\ \text{m}^3/\text{kg s}^2$$
$$= mg, \text{ if } M = \text{mass of earth and } d = \text{radius of earth};$$
$$g = 9{\cdot}8\ \text{m/s}^2 \text{ approximately.}$$

Newton's Laws

1. Everything continues with constant (including zero) velocity unless acted on by a force.

2. A force acting on a body is proportional to, and in the same direction as the acceleration (more precisely, rate of change of momentum) it causes. That is

$$P = ma = d(mv)/dt.$$

3. Whenever a body exerts a force on a second body, the second body exerts an equal and opposite force on the first.

Momentum is defined as mv.

Change of momentum is equal to the impulse causing it. For a constant force

$$mv - mu = tP.$$

For a two-body system, in which there is no external impulse, momentum is conserved: i.e. momentum before impact or explosion $=$ momentum afterwards:

$$m_1v_1 + m_2v_2 = m_1u_1 + m_2u_2.$$

1 Newton (N) gives a mass of 1 kg an acceleration of 1 m/s².

1 kilogram-force (kgf) gives a mass of 1 kg an acceleration of 9·80665 m/s².

16

FURTHER STATISTICS

In this chapter we continue our study of statistics. This is one of the most rapidly advancing subjects in practical mathematics today, and the new ideas here and in the next chapter will go some way to show how it is applied in research and in industry.

We shall develop a measure of spread that is widely used in more advanced work—the standard deviation—and consider a statistic which measures the agreement between two sets of data—the correlation coefficient.

1. MEASURES OF SPREAD

1.1 Properties of a useful measure of spread. The principal use of a measure of spread is to compare the dispersion of one population with the dispersion of another—as, for example, we may say that although the *average* marks awarded in examinations in French and in Mathematics are much the same, the marks in Mathematics are about twice as widely spread. Figure 1 illustrates this.

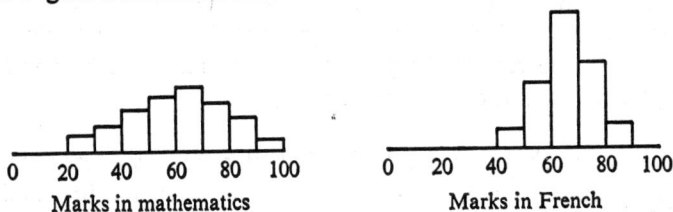

Fig. 1

We shall, however, concentrate upon a single, artificially simple population, and suggest some measures of spread which we might use, to bring out the point of our choice of measure more clearly. Let us take, by way of example, the population

$$P: \quad (3, 5, 6, 6, 7, 9, 13).$$

Now we have already seen how convenient it is, in calculating the mean, to use a working zero and a scale factor, and we must be prepared to think also of related populations, such as

$$A: \quad (63, 65, 66, 66, 67, 69, 73)$$

obtained by a translation of $+60$ on the number line, or

$$B: (300, 500, 600, 600, 700, 900, 1300)$$

obtained by an enlargement of factor 100, or

$$C: \quad (3, 3, 5, 5, 6, 6, 6, 6, 7, 7, 9, 9, 13, 13)$$

obtained by doubling all the frequencies.

Any measure of spread we may choose should have the following properties, if it is to be of practical use to us:

(*a*) Translation along the number line should not affect the measure of spread; thus, whatever number s we use as the measure of the spread of P, s should also be the measure of the spread of A.

(*b*) Enlargement by a factor c should multiply the measure of spread by c; thus, if we use the number s as the measure of spread of P, $100\,s$ should be the measure of the spread of B.

(*c*) Multiplication of all the frequencies by the same factor r should not affect the measure of spread; thus, if we use the number s as the measure, of the spread of P, s should also be the measure of the spread of C.

(*d*) All the members of the population should ideally be taken into account, and yet freaks should not influence the size of the measure unduly.

1.2   The range. Perhaps the simplest measure of spread is the range—the difference between the least and the greatest number of the population.

Now, the range of P is 10; of A, 10; of B, 1000; and of C, 10, so that this measure has properties (*a*), (*b*), and (*c*). But only two of the members have contributed to it, and clearly the largest value, which is fairly widely separated from the rest of the population, and may well be considered freakish, has affected it considerably.

1.3   The inter-sextile range. A slightly better measure—less influenced by freak readings—is the *inter-sextile range*, between the first and fifth sextiles.

The inter-sextile range of P is 4; of A, 4; of B, 400; and of C, 4; once again, (*a*), (*b*), (*c*) are satisfied, although (*d*) is only partially so.

1.4   Deviations. If, however, we calculate the *deviations* of the individual members from some convenient point, often a central value of some sort, we shall still obtain properties (*a*), (*b*), and (*c*); and if we then combine them together in some way, we shall also obtain the property (*d*).

Thus, since the median of P is 6, and the mean 7, the deviations from the median are

$$D: \quad (-3, -1, 0, 0, +1, +3, +7)$$

and from the mean

$$E: \quad (-4, -2, -1, -1, 0, +2, +6).$$

There is no point in averaging these deviations as they stand. (Why not? Try the deviation from the mean first.) There are, however, a number of ways in which we can use them to find a satisfactory measure of spread.

1.5   Mean absolute deviation. If we neglect the signs of these deviations, and make them into absolute deviations (that is to say, distances) from the central value, instead of the average deviation we can calculate that the *mean absolute deviation* from the median is $2\frac{1}{7}$, and that the *mean absolute deviation* from the mean is $2\frac{2}{7}$.

Since the median of D, by definition, is 0 (why?), and since the dispersions of the populations P, A, C, D, and E, are precisely the same, because they are all related by translations on the number line, the mean absolute deviation of each of these populations from its median is $2\frac{1}{7}$, while for B it is $214\frac{2}{7}$.

By a similar argument, starting from E, the mean absolute deviation from the mean is $2\frac{2}{7}$ for populations P, A, C, D, E, and so for population B it is $228\frac{4}{7}$.

Property (d) is reasonably well satisfied by this measure; but it has a serious disadvantage.

Suppose that we want to use a working zero to calculate the mean absolute deviation from the mean. Let us try using 0, 2, 4, 6, and 8 as working zeros, and see how the mean absolute deviation from each of these numbers differs from the mean absolute deviation from the mean. We shall want to see if there is a straightforward method of passing from the mean absolute deviation from a working zero to the mean absolute deviation from the mean.

The calculations are straightforward, and the reader should verify that the answers are as follows:

Working zero	Mean absolute deviation from working zero	To find mean absolute deviation from mean
0	7	Subtract $4\frac{5}{7}$
2	5	Subtract $2\frac{5}{7}$
4	$3\frac{2}{7}$	Subtract 1
6	$2\frac{1}{7}$	Add $\frac{4}{7}$
8	$2\frac{5}{7}$	Subtract $\frac{3}{7}$

It is fairly clear that no obvious pattern emerges. The difficulty is that we cannot 'correct' to the mean—or to the median—without referring to the individual members of the population; and this makes calculations with large populations tedious.

1.6 Standard deviation. Oddly enough, if instead of simply suppressing the signs of the deviations, we *square* the deviations, the correction can be very simply made.

Consider the *mean squared deviation* from each of the natural numbers from 0 to 8. Again, the reader should verify that the mean squared deviation from each of these numbers is as shown, and that this table gives the right corrections to alter them into the mean squared deviations from the mean.

Working zero	Mean squared deviation from working zero	To find mean squared deviation from mean
0	$57\tfrac{5}{9}$	Subtract 49
1	$44\tfrac{5}{9}$	Subtract 36
2	$33\tfrac{5}{9}$	Subtract 25
3	$24\tfrac{5}{9}$	Subtract 16
4	$17\tfrac{5}{9}$	Subtract 9
5	$12\tfrac{5}{9}$	Subtract 4
6	$9\tfrac{5}{9}$	Subtract 1
7	$8\tfrac{5}{9}$	—
8	$9\tfrac{5}{9}$	Subtract 1

Here, a very remarkable pattern emerges; the 'correction' to the mean is simply the square of the difference between the working zero and the mean. This is always true, though we will not prove it here (compare Exercise B, Question 16), and it gives us, unexpectedly, property (*e*):

(*e*) it is easy to use a working zero to calculate this measure.

But, in the process, we have lost (temporarily) property (*b*). As the reader will easily verify this measure for population B is not, as we require, $8\tfrac{5}{9} \times 100$, but $88571\tfrac{1}{3}$ $(= 8\tfrac{5}{9} \times 10^4)$. The reason for this may be better understood from the following argument. If the members of the population are the lengths in *metres* of a number of objects, the deviations have been measured in metres, and the squared deviations in *square metres*. If, therefore, we re-write the data in *centimetres*, as in population B, we multiply the mean squared deviation by 10^4, because it is now measured in *square centimetres*. We rectify this by taking the *square root* of this as our final measure, and property (*b*) is restored.

This measure is the *standard deviation*, much used in advanced work, partly because it is so easy to calculate; and the complicated process which leads up to it is admirably summed up in its alternative name, the *root-mean-squared deviation from the mean*.

Example 1. Find the mean and standard deviation of the population (13, 15, 16, 17, 19), and sketch a histogram to show their appearance on the number-line.

We take a working zero k at 16, and a class interval c of 1, and we observe that the number of members of the population is 5. $k = 16$, $c = 1$, $N = 5$. Then $\Sigma t = 0$, so that the mean is 0, and the t's are deviations from the mean.

The mean squared deviation from the mean, s_t^2, is therefore

$$\Sigma t^2/N = 20/5 = 4;$$

the standard deviation is therefore 2.

x	t	t^2
13	-3	9
15	-1	1
16	0	0
17	$+1$	1
19	$+3$	9
Σ	0	20

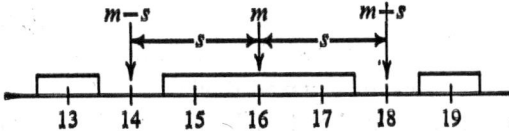

Fig. 2

Exercise A

1. Repeat the working of Sections 1.5 and 1.6 for the population (3, 5, 6, 7, 9). Compare your answers with those of Example 1.

Find the mean and standard deviation of each of the populations given in Questions 2–6, and sketch histograms to show their appearance on the number-line.

2. (2, 4, 6, 6, 12).   **3.** (3, 4, 5, 6, 7, 8, 9).

4. (8, 11, 14, 17, 20, 23, 26).
(Compare the answers to Questions 2, 3 and 4 with the answer to Example 1 and comment on any similarities between them.)

5. The heights of a form of 13-year-old boys: 1·5 m, 1·8 m, 1·7 m, 1·55 m, 1·65 m, 1·85 m, 1·6 m, 1·5 m, 1·5 m, 1·65 m.

6. The heights of a form of 13-year-old girls: 1·55 m, 1·8 m, 1·65 m, 1·65 m, 1·6 m, 1·7 m, 1·55 m, 1·55 m, 1·6 m, 1·65 m.
Comment on the connection between the answers to Questions 5 and 6.

2. CALCULATION OF THE STANDARD DEVIATION

2.1 Notation and arrangement of work. In this chapter, we follow the tabular arrangement of calculations used in earlier books, but we adopt the notation of more advanced work, as being more suitable at this level. A summary of this notation is given here.

Actual quantity considered	x
Frequency of this quantity	f
Working zero	k
Class interval	c
Transformed variable (that is, $(x-k)/c$)	t
Number in sample	$N = \Sigma f$

Mean
$$m = \frac{\Sigma xf}{\Sigma f} = k + c\frac{\Sigma tf}{\Sigma f}$$

Standard deviation (if ambiguity is possible, this may be written s_x or s_t, for example.)
$$s = \sqrt{\left\{\frac{\Sigma x^2 f}{\Sigma f} - \left(\frac{\Sigma xf}{\Sigma f}\right)^2\right\}}$$
$$= c\sqrt{\left\{\frac{\Sigma t^2 f}{\Sigma f} - \left(\frac{\Sigma tf}{\Sigma f}\right)^2\right\}}$$
(see below).

2.2   Correcting the mean squared deviation. In general the mean is not itself a convenient number to take as a working zero (Exercise A carefully avoided this difficulty). But in Section 1.6 we found a method of 'correcting' from the working zero to the mean, and this result can easily be expressed in words like this:

> To find the mean squared deviation from the mean, take the mean squared deviation from the working zero, and subtract the square of the difference between the working zero and the mean.

We can summarize this by saying that by definition the *variance*, or square of the standard deviation, is given by

$$s^2 = \frac{\Sigma(x-m)^2}{N},$$

but in practice we generally calculate it by writing d as the deviation of each reading from the working zero and using

$$s^2 = \frac{\Sigma d^2}{N} - \left(\frac{\Sigma d}{N}\right)^2.$$

This is, of course, because $m - k = \Sigma d / N$. It is easy to see how this is extended to the general formula quoted in Section 2.1.

2.3   Worked examples.

Example 2. Find the mean and standard deviation of the lengths given in the table (which have been arranged in order of magnitude):

x	t	t^2
61	$-\ 8$	64
$62\frac{1}{2}$	$-\ 2$	4
$63\frac{1}{4}$	$+\ 1$	1
$63\frac{3}{4}$	$+\ 3$	9
$64\frac{1}{2}$	$+\ 6$	36
$65\frac{1}{2}$	$+10$	100
Σ	$+10$	214

Take $k = 63$ cm, $c = \frac{1}{4}$ cm, $N = 6$. Then $\Sigma t = 10$ and $\Sigma t^2 = 214$, so that

$$m_t = \tfrac{10}{6} \simeq 1.67,$$

and

$$s_t^2 = \frac{\Sigma t^2}{N} - \left(\frac{\Sigma t}{N}\right)^2$$

$$\simeq 35.7 - 2.8 = 32.9,$$

so that $s_t \simeq 5.7$. Transforming back from t to x, and rounding the results to give a reasonable degree of accuracy,

$$m = 63 \text{ cm} + \tfrac{1}{4} \text{ cm} \times 1.67 = 63.4 \text{ cm}; \quad s = \tfrac{1}{4} \text{ cm} \times 5.73 = 1.4 \text{ cm}.$$

Example 3. Four coins are tossed, and the number of heads obtained is written down. This is done 60 times, and the results are set out in the table below. Find the mean and standard deviation. (It can be shown, as in Chapter 17, that the theoretical result for *fair* coins gives a mean of 2 and a standard deviation of 1 exactly.)

Take $k = 2$, $c = 1$.

x	f	t	tf	t^2f
4	5	$+2$	$+10$	20
3	14	$+1$	$+14$	14
2	23	0	0	0
1	15	-1	-15	15
0	3	-2	$-\ 6$	12
			$+24-21$	
Σ	60	—	$+3$	61

Then $\qquad \Sigma f = 60, \quad \Sigma tf = +3, \quad \Sigma t^2 f = 61,$

so

$$\frac{\Sigma tf}{\Sigma f} = \frac{3}{60} = 0.05,$$

and

$$s_t = \sqrt{\left\{\frac{\Sigma t^2 f}{\Sigma f} - \left(\frac{\Sigma tf}{\Sigma f}\right)^2\right\}} = \sqrt{\left(\frac{61}{60} - \frac{9}{3600}\right)}$$

$$\simeq 1.007.$$

Transforming from t to x, and rounding to suitable accuracy, $m = 2.05$, $s = 1.01$.

Example 4. The frequencies of marks in a Physics examination are shown in the table. Estimate the mean and the standard deviation.

(Note: since the marks are grouped the answers can only be approximate.)

Each group is represented by its mid-mark, x. Take $k = 55.5$; $c = 10$.

Range	Mid-mark x	f	t	tf	t^2f
11–20	15·5	30	−4	−120	480
21–	25·5	60	−3	−180	540
31–	35·5	220	−2	−440	880
41–	45·5	540	−1	−540	540
51–	55·5	490	0	0	0
61–	65·5	310	+1	+310	310
71–	75·5	180	+2	+360	720
81–	85·5	110	+3	+330	990
91–100	95·5	60	+4	+240	960
				+1240−1280	
Σ	—	2000	—	−40	+5420

Then

$$\Sigma f = 2000,$$

$$\Sigma tf = -40,$$

$$\Sigma t^2 f = 5420,$$

so

$$\frac{\Sigma tf}{\Sigma f} = -0.02;$$

and

$$s_t = \sqrt{\left\{ \frac{\Sigma t^2 f}{\Sigma f} - \left(\frac{\Sigma tf}{\Sigma f} \right)^2 \right\}} = \sqrt{\left\{ \frac{5420}{2000} - \left(\frac{-40}{2000} \right)^2 \right\}}$$

$$= 1.65.$$

Transforming from t to x, we have

$$m = 55.5 - 10 \times 0.02 = 55.3;$$

and

$$s_x = 10 \times 1.65 = 16.5.$$

2.4 Checking calculations.

1. The *inter-sextile range* is approximately equal to two S.D.'s, when the distribution is Normal (that is, of the typical bell-shape of Chapter 17).

This is a very simple check, and it can quickly be applied. It will usually show up any major error.

Check: I.S.R. = $2s$ roughly.

2. *Charlier's checks.* For questions such as Example 4, when there are many figures to check and mistakes can easily be overlooked, it is useful to add two columns to the table and to calculate $\Sigma(t+1)f$ and $\Sigma t(t+1)f$.

362

Then we check these new figures against the old, thus:

$$\Sigma(t+1)f = \Sigma tf + \Sigma f \quad \text{and} \quad \Sigma t(t+1)f = \Sigma t^2 f + \Sigma tf.$$

All these totals are available in the enlarged table, and this sort of searching check, associated with the name of M. Charlier, is important.

Example 4 (*continued*). To show the checks. To the table above add

$t+1$	$(t+1)f$	$t(t+1)f$
-3	-90	360
-2	-120	360
-1	-220	440
0	0	0
$+1$	$+490$	0
$+2$	$+620$	620
$+3$	$+540$	1080
$+4$	$+440$	1320
$+5$	$+300$	1200
	$+2390-430$	
Σ	$+1960$	5380

Charlier's checks:

$$\Sigma(t+1)f = \Sigma tf + \Sigma f,$$
$$1960 = 2000 - 40\checkmark;$$
$$\Sigma t(t+1)f = \Sigma t^2 f + \Sigma tf,$$
$$5380 = 5420 + (-40)\checkmark;$$
$$\text{I.S.R.} = 72 - 41 = 31,$$
$$2s = 33\checkmark.$$

Both checks are satisfactory; Charlier's checks should be exact, but the I.S.R. will always be only approximate.

Exercise B

Find the mean and standard deviation of the readings given in Questions 1–10.

1. The number of splashes observed in 20 m intervals with centres at 1020, 1040, ..., 1140 m when a machine gun was fired over water: 2, 7, 25, 36, 19, 10, 1.

2. Numbers of heads in 16 tosses of four coins (theoretical frequencies): 4 heads, 1; 3 heads, 4; 2 heads, 6; 1 head, 4; 0 heads, 1.

3. The time taken by a good timekeeper to react, in 1/100 seconds, in 500 experiments:

Time (seconds)	0·15	0·16	0·17	0·18	0·19	0·20	0·21	0·22	0·23	0·24	0·25
Frequency	2	12	25	50	92	136	98	46	23	14	2

4. Batsman's scores: compare and comment:

Player A	93	57	0	8	60	8	4	10
Player B	24	34	43	33	18	38	20	30

5. Lap times (in seconds) for standard saloon cars are distributed as follows:

Class mid-marks	125	127·5	130	132·5	135	137·5	140	142·5
Frequency	2	4	8	21	37	18	6	4

6. Number of defectives in 1296 samples of 4, drawn from a population in which $\frac{1}{6}$ are in fact defective (theoretical result):

No. of defectives	4	3	2	1	0
No. of samples	1	20	150	500	625

7. The scores in the thirty-six possible outcomes of a single throw of two dice.

8. The amount of stoppages (in francs) for punishment in the Foreign Legion.

Amount	0	13·5	27	40·5	54
Frequency	148	87	29	4	1

9. The I.Q.'s of a group of boys.

I.Q.	106–110	111–115	116–120	121–125	126–130	131–135
Frequency	6	15	16	8	3	2

10. Time taken to pick a basket of fruit of a given size, observed over a complete day's work (30 pickers):

Time (min)	20–25	25–30	30–35	35–40
Frequency	143	193	126	52

11. The *coefficient of variation* can be defined as s.d./mean (sometimes expressed as a percentage). Use some of the questions above to see what it measures, and discuss its advantages and disadvantages.

12. In which of the questions above does half the inter-sextile range (that is, half the difference between the first and fifth sextiles) *not* give a good estimate of the s.d.? For what type of distribution is this estimate useful?

13. Collect data on your form's marks in several subjects, find the mean and s.d. in each set of subject marks. If these marks are added together, compare the results of (*a*) adding them as they stand, and (*b*) adding them after giving them all the same approximate s.d.

14. Seven boys' marks in Geography and History were:

| Geography | 30 | 52 | 60 | 100 | 78 | 39 | 82 |
| History | 75 | 90 | 78 | 93 | 85 | 84 | 97 |

Calculate the mean and standard deviation of the marks obtained in each subject. Explain which you consider to be the better mark—78 in Geography or 85 in History.

Plot the results as ordered pairs (30, 75), (52, 90), ..., (82, 97). What does the scatter suggest? Estimate the fairest mark to give a boy who missed the Geography exam, but scored 79 in History.

15. Prove that the mean absolute deviation from the median is never greater than the mean absolute deviation from any other number.

16. Prove that the population (a, b, c, y, z), for which $N = 5$, has the property defined in Section 2.2. Can you use the notation to extend this proof to a population of N members?

* Extend this result further to prove the result quoted in Section 2.1.

3. RANK CORRELATION

You will have heard many people talking about a 'correlation' between two lots of figures. The word is often used loosely, but like most of the terms of statistics it is precise, and must be defined and used carefully. There is a correlation between the amount of alcohol in people's bloodstreams and the number of accidents they have. There may be a correlation between mathematical and musical ability; there is a correlation in the opposite sense between the ages at which babies walk, and the ages at which they talk. The question is how much importance we can attach to an apparent connection between the two lots of figures.

We need a measure of this connection which can be exactly defined and whose meaning can be objectively evaluated. The conventional method of measuring such agreement is to calculate a *correlation coefficient*; that is, a number which ranges from $+1$ for perfect agreement, through 0, for no real connection at all, to -1 for complete reversal.

All correlation coefficients have this same range of values and are interpreted in much the same way. If you can understand and interpret one, then you will find the others fairly easy to use when you come across them. The full correlation coefficient is beyond the scope of this book, and, of the *rank correlation coefficients* we shall investigate Kendall's (1948), rather than Spearman's (1906), as it is a rather simpler introduction to the subject.

There are two advantages of rank correlation, that is, working with ranks and not with actual numerical scores. First, the calculation is very simple, since it uses only small whole numbers. Secondly, the method can be used where it is not possible to give a precise numerical value to each quantity, but it is possible to place any pair in order of merit. The method sometimes has disadvantages—ranking figures may impose on them a distortion, because the differences between adjacent terms are all made equal.

3.1 Kendall's Rank Correlation Coefficient τ.

Example 5. Find Kendall's Rank Correlation Coefficient for the agreement between two judges in a Miss World contest:

	Miss America	Miss Bulgaria	Miss China	Miss Denmark	Miss England	Miss Finland
Peter	2	4	3	6	1	5
Quong	4	3	1	5	2	6

Kendall's Rank Correlation Coefficient—hereafter called K.R.C.C. or τ—is calculated by considering how many of the judges' decisions on the 15 pairs of girls agree.

We build up a score like this. Consider a particular pair of girls. If both judges agree on the order in which they place them, $+1$ is added to the score. If the judges disagree, -1 is added to the score. If either judge ranks the two girls equal, the score is left unchanged.

There are, in this case, fifteen contributions to the score; if the judges agree over each pair, the score is $+15$; if they disagree over each pair, it is -15. The *average* of these contributions is Kendall's Rank Correlation Coefficient, which therefore has the range of values, -1 to $+1$, suggested above. In this example the method is explained at length and you can see best how to set out the work.

First, re-write the table so that the girls are listed in judge P's order of merit, as in the table below. We now need to consider only the order in judge Q's row.

Secondly, start with the left-hand number in Q's row, that is, 2. Now, considering only Q's row, count $+1$ for every number to the right which is greater than 2, and -1 for every number to the right which is less than 2.

Thirdly, do the same with the second number, 4, remembering to consider only numbers to its right. Carry on until the 6 is reached, recording the scores underneath. You can now find the total score.

	E	A	C	B	F	D	
P	1	2	3	4	5	6	
Q	2	4	1	3	6	5	
	$+4$	$+2$	$+3$	$+2$	0	—	
	-1	-2	0	0	-1	—	
	$+3$	0	$+3$	$+2$	-1	—	Total $+7$

You have in fact considered the agreement or disagreement for each of the fifteen pairs. You should convince yourself of this. For example, if you take the pair *EA*, you can see that judge *P* preferred *E*, and so did judge *Q*. The score is therefore correctly increased by $+1$. If, however, we take the pair *AB*, *P* preferred *A*, but *Q* preferred *B*. The score was therefore increased correctly by -1, because in *Q*'s row we saw that 3 was to the right of 4.

The total number of pairs, which is also the largest possible number of agreements, can be counted in the same way. Imagine that there is perfect correlation, that is, that the judges placed all six girls in the same order. Then you can count as follows:

1	2	3	4	5	6	
1	2	3	4	5	6	
$+5$	$+4$	$+3$	$+2$	$+1$	—	Total 15

In this example, the average score is 7/15, and this, by definition, is Kendall's Rank Correlation Coefficient. It is usually quoted as a decimal: $\tau = 0.47$.

You should be able to see that perfect agreement will give K.R.C.C. as 15/15, that is, 1, and complete reversal will give $-15/15$, that is, -1; the coefficient therefore fulfils the basic criterion.

Example 6. Seven children are ranked by their I.Q.'s and by the number of O-levels they pass. Find K.R.C.C. for these two orders.

By I.Q.	1	2	3	4	5	6	7	
By O-level	2	$3\frac{1}{2}$	$3\frac{1}{2}$	1	6	5	7	
	$+4$	$+2$	$+2$	$+3$	0	$+1$	—	Total 12
Largest possible	$+6$	$+5$	$+4$	$+3$	$+2$	$+1$	—	Total 21

Therefore $\tau = 12/21 = 0.57$.

This example shows what happens if one pair is ranked equal by one of the judges. It is easy to deal with this if the equality occurs in the second row. If it occurs only in the top row, the rows can be switched. If it occurs in both rows, *no* score must be added as a result of the entries ranked equal in the top row.

3.2 The formula for K.R.C.C.

The largest possible score can also be found from the formula $\frac{1}{2}n(n-1)$, where n is the number ranked.

In Example 6, $n = 6$, so that $\frac{1}{2}n(n-1) = 15$; and in Example 7, $n = 7$, so that $\frac{1}{2}n(n-1) = 21$, as indeed we found by the methods suggested above.

K.R.C.C. is cumbersome to express as a formula, but you may find it easy to remember as

$$\frac{\text{number of agreements} - \text{number of disagreements}}{\frac{1}{2}n(n-1)}.$$

Exercise C

Calculate τ (Kendall's Rank Correlation Coefficient), in the following questions. Keep your results for use in the next exercise.

1. Five stop-watches, of different prices, are graded for accuracy.

Price (grade)	1	2	3	4	5
Accuracy (grade)	1	3	4	2	5

(Answer: $\tau = 0.6$)

2. Five typists are tested for speed and accuracy.

Speed (w.p.m.)	67	63	61	58	57
Accuracy (%)	87	85	91	96	93

(Rank the figures first)

3. Six different soaps are tested for height of lather in a test tube:

Pence for 50 g	10	7	4	5	8	12
Lather (mm)	21	22	16	18	24	17

(Fictitious)

4. Two firms grade ten brands of margarine for 'buttery flavour'.

Brand	Q	R	S	T	U	V	W	X	Y	Z
Firm A (grade)	10	5	6	1	4	2	3	8	9	7
Firm B (grade)	10	2	8	3	6	1	5	9	7	4

Is the evidence better among any smaller sets of brands?

5.

							More than
(a) Size of limpets in cm (class midmark)	Up to 2	2·1	2·3	2·5	2·7	2·9	3
(b) Mean height above low water mark (in metres)	2·3	2·1	3·4	4·1	4·2	5·0	7·1
(c) Slope of shell (10 × ht/mean radius)	3·4	3·0	3·0	4·7	5·0	5·2	5·8

Find three values of τ: τ_{ab}, τ_{bc}, τ_{ca}.

368

4. SIGNIFICANCE

In Example 5, the number 7/15, by itself, tells us practically nothing at all. It does, however, give a numerical value which can be compared with the number reached after other similar calculations. As a matter of fact, the agreement would have been very much more impressive if this fraction had come from an ordering of 25 places, instead of only 6. However, there are two useful questions that we may ask.

I. How unlikely is it that, if one or both of the rankings had been made purely at random, we should achieve a score as high as this or higher?

II. If it is most unlikely that so high a coefficient is the result of chance, can we tell if there is a cause and effect relationship between the two lots of figures?

4.1 Interpretation. To take the second question first: it is hard to give a satisfying answer here. For example, there are strong positive correlations, over the post-war years, between each pair of the following five lots of figures:

> the number of T.V. sets licensed;
> the incidence of juvenile delinquency;
> the number of refrigerators owned;
> the number of crime films of T.V.;
> the duty on a litre of petrol.

Of course, no serious statistician would ever make the mistake of supposing that a high correlation coefficient necessarily implies that there is a cause and effect relationship. It would, for example, be ridiculous to suggest that the increasing number of refrigerators owned is a cause of the increasing amount of juvenile delinquency. Analyse and quantify as much as you can, to reduce the amount of guessing that you do, but remember that a healthy scepticism is the most important item in the statistician's equipment.

4.2 Level of significance. The first point, however, is more easily answered. The larger the sample and the higher the coefficient, the more likely it is that there is a connection between the items.

A coefficient as big as (or bigger than) the 7/15 (0·47) of Example 5 would occur about 14 % of the time by chance. You are not required to work this out in every case; statisticians use tables or graphs and just look it up, and so shall we; but the sort of calculation involved runs like this.

There are altogether 720 possible ways in which the 6 girls could have

369

been arranged in order. The frequencies of the possible relevant scores (ignoring equalities) are approximately Normal† and work out as

+15	+13	+11	+9	+7	+5	+3	+1
1	5	14	29	49	71	90	101

−1	−3	−5	−7	−9	−11	−13	−15
101	90	71	49	29	14	5	1

If each of these orders is equally likely, as would be the case if in fact there was no correlation (that is, if one judge had no skill or was using a random process) a score of +7 or better would occur 98 times out of the 720, roughly 14 % of the time, which is by no means an unlikely event. Statisticians start to take notice when the frequency is as low as 5 %, and grow progressively more interested when it drops to 1 % or 0·1 %. You can think of these last two as giving a probability of an accidental occurrence of only 1/100 and 1/1000 respectively, and then you will see how strong the connection is.

4.3   Using the graph. The three levels commonly used, 5 %, 1 % and 0·1 %, have contours as drawn on the chart. (See Figure 3.) The number in the sample, i.e. the number ranked, is shown on the horizontal axis, and the size of τ on the vertical axis. The point opposite 6 on the 'number ranked' axis and 0·47 on the τ axis is indicated ⊙. This is in the '*not significant*' part of the graph.

We should therefore sum up our findings in ordinary language as: '$\tau = 0·47$, which does not show a significant degree of agreement between the judges'.

Example 6, with $\tau = 0·57$ for 7 ranked items, gives the point marked ⊗, and the result is therefore significant at the 5 % level.

Exercise D

1–5. Estimate the significance of the values of τ obtained in Exercise C (Questions 1–5) and express your conclusions in ordinary language.

Calculate τ, estimate its significance, and comment on the results in Questions 6–9.

6. Car and motor-cycle licences compared with bus and coach journeys:

Year	1952	1953	1954	1955	1956	1957	1958	1959	1960	1961
Licences ($\times 10^6$)	3·5	3·8	4·2	4·8	5·2	5·7	6·1	6·7	7·4	7·8
Journeys ($\times 10^9$)	16·3	16·1	15·9	15·9	15·4	14·7	13·8	13·9	13·7	13·4

Why is the coefficient not very helpful in this case?

† This is discussed at length in the next chapter.

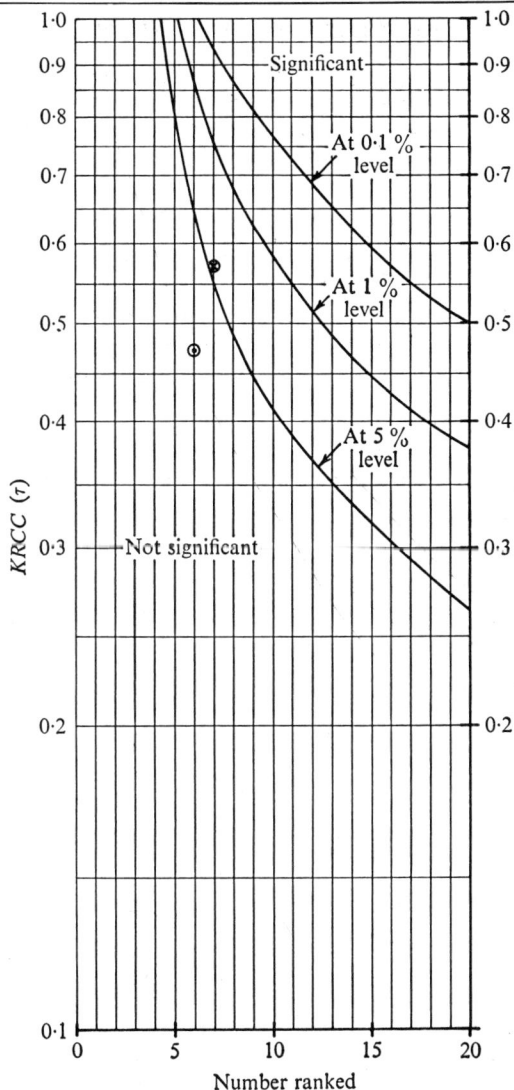

Fig. 3. Significance levels of τ (Kendall's Rank Correlation Coefficient).

7. A set of boys, listed in order of merit in the summer term, scored the following O-level marks:

Order	1	2	3	4	5	6	7	8	9	10
Marks	74	61	61	81	53	76	68	85	61	70
Order	11	12	13	14	15	16	17	18	19	20
Marks	56	78	58	64	72	49	52	64	68	62

Can you make any general comment about the comparison? Which boys seem to have been the least consistent?

8. Output of farm crops (at 1955 prices) compared with rainfall from June to August:

Year	1954	1955	1956	1957	1958	1959	1960	1961	1962
Output (£m)	274	251	252	273	248	248	288	301	284
Rain (cm)	22·5	30	20	32·5	25	30	15	27·5	17·5

Can you think of a set of figures which you would expect to agree better with the output figures?

9. In a competition building up cumulatively over six weeks, the scores were as follows:

House

Week	A	B	C	D	E	F	G	H	I	K
1	29	46	59	32	23	12	23	32	26	26
2	64	61	84	61	56	35	50	61	46	41
3	92	89	90	93	84	59	74	72	71	60
4	108	99	102	100	94	67	81	78	73	64
5	126	100	113	108	99	82	93	83	75	75
6	140	118	117	116	100	101	98	87	84	83

Scores given (cumulative)

Calculate in which week the agreement of the order with the final week first became significant.

*10. Write out all the possible orders for four grades, and their K.R.C.C.'s when compared with the order 1234. Hence, taking each order as equally likely, give the probability of each score.

*11. In Question 10, find the effect of inserting a '5' in each possible place in turn, and hence repeat Question 10 for five grades. Show how to extend the process to six grades, and confirm the working of Section 4.2.

*12. Show that in an order in which two are ranked equal, the score will be the average of the two possible scores if they were arbitrarily separated.

*13. Give instructions for writing down the score in Question 7 without writing down anything else before it, and without converting the figures of the second row to ranks.

*14. Is it possible to calculate the score:

(a) by comparing each row in turn with the order $123...n$ and combining the answers;

(b) by considering the differences in rank awarded by the two judges;

(c) by any other method?

In each case, give a reason if it is possible, or an example to show that it is not possible.

*15. Collect marks for people in your form for four subjects. Rank them in these subjects and work out the six correlations. Add the marks in a fair way (i.e. make sure that the standard deviations or the ranges are about equal), rank this total and compare the four new coefficients you get with the others. Display your results in tabular form and add your comments.

17

FURTHER PROBABILITY

1. PROBABILITY PATTERNS

1.1 Rectangular histogram. We define a fair die by saying that, when it is rolled, each number from 1 to 6 is equally likely to turn up. We then say that each number has a probability of 1/6.

The histogram of possibilities for a fair die is a simple rectangle, illustrated in Figure 1.

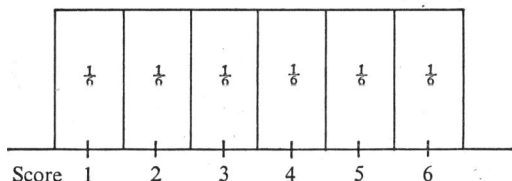

Fig. 1

1.2 Triangular histogram. To extend this idea to two dice involves a little more thought. We can get any total from 2 to 12, but the totals are not all equally likely. To help us to see the different probabilities we draw a diagram (Figure 2) which supposes that one of the dice is red and the other

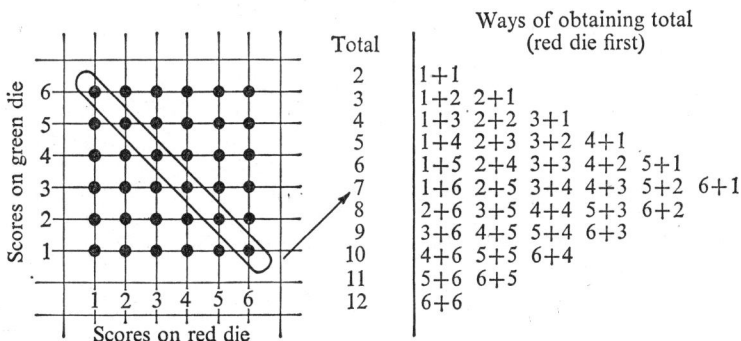

Total	Ways of obtaining total (red die first)
2	1+1
3	1+2 2+1
4	1+3 2+2 3+1
5	1+4 2+3 3+2 4+1
6	1+5 2+4 3+3 4+2 5+1
7	1+6 2+5 3+4 4+3 5+2 6+1
8	2+6 3+5 4+4 5+3 6+2
9	3+6 4+5 5+4 6+3
10	4+6 5+5 6+4
11	5+6 6+5
12	6+6

Fig. 2

is green, and we see the thirty-six possible throws, and the scores that may result from each. Thus, for example, we see that we can get a total of 7 in six different ways.

373

Since the possibility space is of size 36, and of these thirty-six equally likely outcomes six give a total of 7, we may say that the probability of obtaining a total of 7 is 6/36, that is, 1/6.

Similarly, the probability of getting a 4 is 3/36, that is to say, 1/12.

The histogram, illustrated in Figure 3, is roughly triangular.

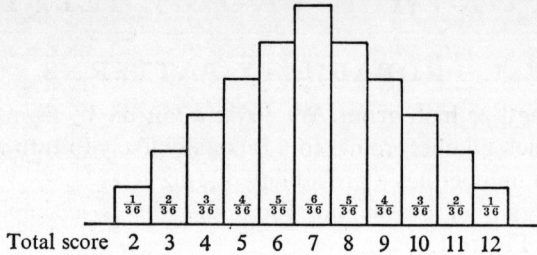

| | $\frac{1}{36}$ | $\frac{2}{36}$ | $\frac{3}{36}$ | $\frac{4}{36}$ | $\frac{5}{36}$ | $\frac{6}{36}$ | $\frac{5}{36}$ | $\frac{4}{36}$ | $\frac{3}{36}$ | $\frac{2}{36}$ | $\frac{1}{36}$ |

Total score   2   3   4   5   6   7   8   9   10   11   12

Fig. 3

1.3   Pascal's triangle.

Example 1. What is the probability of 0, 1, 2, ..., m heads occurring when a coin is tossed m times?

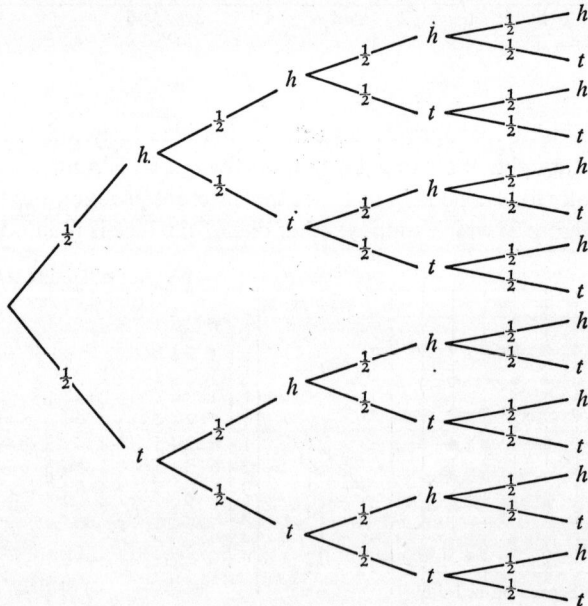

Fig. 4

We assume that the coin is symmetrical, and that there is no possibility of its standing on its edge. This means that the probability along each part of each branch is $\frac{1}{2}$. Figure 4 shows a tree diagram which illustrates the possibilities when the coin is tossed four times. We then see that the probability along each branch is $\frac{1}{2} \times \frac{1}{2} \times \frac{1}{2} \times \frac{1}{2}$, that is $\frac{1}{16}$. The problem is therefore reduced to finding out how many branches there are which give each of the possible outcomes.

The branches can be grouped as follows:

0 head, {tttt}	1 outcome
1 head, {httt, thtt, ttht, ttth}	4 outcomes
2 heads, {hhtt, htht, htth, thht, thth, tthh}	6 outcomes
3 heads, {hhht, hhth, hthh, thhh}	4 outcomes
4 heads, {hhhh}	1 outcome

The reader should write out similar tables for 1, 2, 3, and 5 tosses.

A table can then be prepared with the appropriate probabilities entered as below:

Probability of specified number of heads

Tosses	0	1	2	3	4	5
1	$\frac{1}{2}$	$\frac{1}{2}$	—	—	—	—
2	$\frac{1}{4}$	$\frac{2}{4}$	$\frac{1}{4}$	—	—	—
3	$\frac{1}{8}$	$\frac{3}{8}$	$\frac{3}{8}$	$\frac{1}{8}$	—	—
4	$\frac{1}{16}$	$\frac{4}{16}$	$\frac{6}{16}$	$\frac{4}{16}$	$\frac{1}{16}$	—
5	$\frac{1}{32}$	$\frac{5}{32}$	$\frac{10}{32}$	$\frac{10}{32}$	$\frac{5}{32}$	$\frac{1}{32}$

and so on

There are two interesting patterns in this table. The denominators (lower lines) are simply the appropriate powers of 2, and the numerators (upper lines) are all very closely connected with the numerators in the previous row.

In fact, the pattern of the numerators is the beginning of Pascal's triangle

$$\begin{array}{ccccccccc}
& & & & 1 & & & & \\
& & & 1 & & 1 & & & \\
& & 1 & & 2 & & 1 & & \\
& 1 & & 3 & & 3 & & 1 & \\
1 & & 4 & & 6 & & 4 & & 1 \\
\end{array}$$

This triangle can be extended indefinitely by adding together each pair of adjacent terms and writing their sum in the space below them. Complete the next four lines for yourself.

The tree diagram in fact shows why this adding process builds the triangle up for us; for if we know that *three* branches of the tree-diagram for three tosses have just one head, and *three* have just two heads, we can see at once why *six* branches of the tree diagram for four tosses have just two heads.

375

The histograms of some of these probability patterns are shown in Figure 5, together with an idealization of them, the Normal curve, which is the shape that they approach if we increase n, but arrange the scale so that the area under the curve is kept constant.

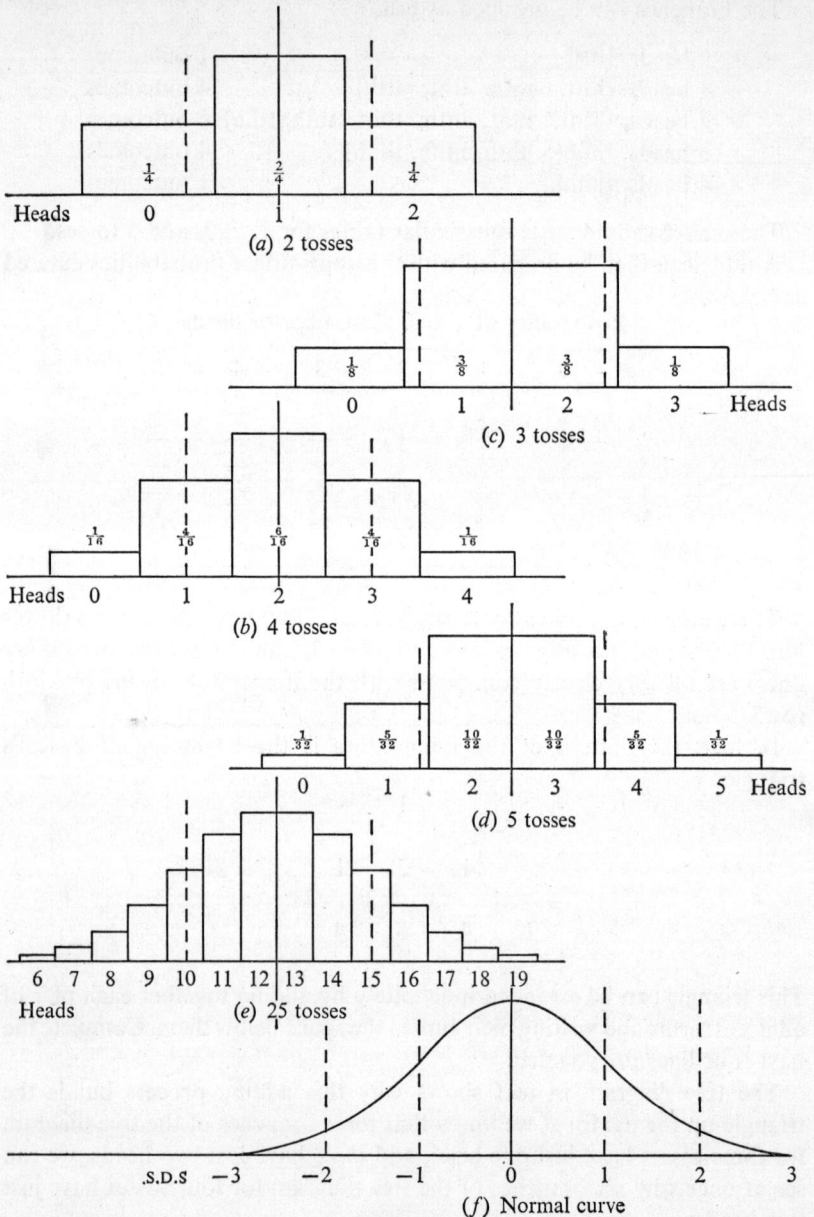

(a) 2 tosses

(c) 3 tosses

(b) 4 tosses

(d) 5 tosses

(e) 25 tosses

(f) Normal curve

Fig. 5

1.4   Repeated trials in general.

Example 2. The Doghouse missile is so unreliable (due to electrical faults, human error, and so on) that its chance of success is only 1/6. Find the probabilities of each possible number of successes in a firing of four of these missiles.

Write p for the probability of success, that is, 1/6, and q for the probability of failure, 5/6, to make the working easier to appreciate, and to make later generalizations possible. Notice that $p+q = 1$.

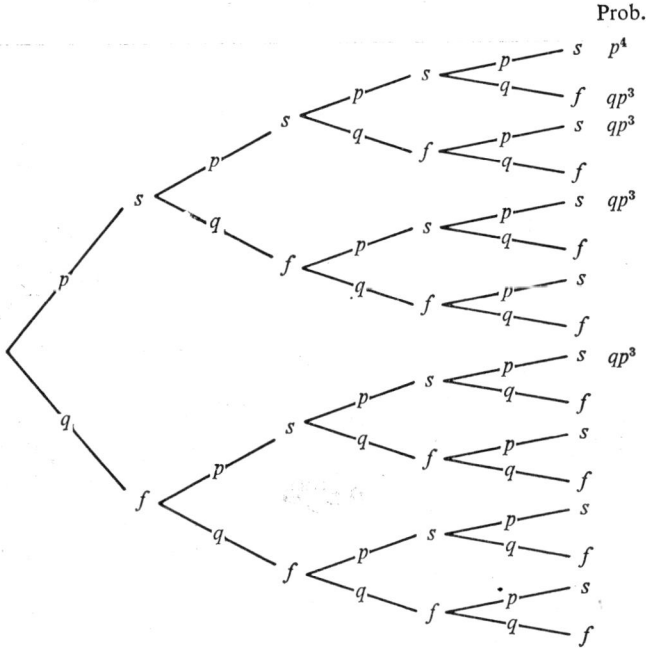

Fig. 6

It is easy to see at once from the tree diagram that the probability of *four* successes is p^4. But the probability of *three* successes involves two separate problems, because there are several branches which represent just three successes and one failure. Each of these branches is labelled in Figure 6, and the probability of the outcome represented by each of these complete branches is qp^3. Moreover, there are four of these branches, exactly as in the previous example, so that the probability of just three successes is $4qp^3$, that is, 20/1296.

The probability of i successes may be calculated in just the same way. Each branch which represents i successes, and therefore $4-i$ failures, will have the probability $q^{4-i}p^i$. Moreover, Pascal's triangle tells us how many such branches there are, as in Example 1.

We therefore find, for this particular instance, that the results may be tabulated like this:

Number of successes	0	1	2	3	4
Probability (from tree diagram)	q^4	$4q^3p$	$6q^2p^2$	$4qp^3$	p^4
Probability (in this case)	$(\frac{5}{6})^4$	$4(\frac{5}{6})^3 (\frac{1}{6})^1$	$6(\frac{5}{6})^2 (\frac{1}{6})^2$	$4(\frac{5}{6})^1 (\frac{1}{6})^3$	$(\frac{1}{6})^4$
Probability (simplified)	$\frac{625}{1296}$	$\frac{500}{1296}$	$\frac{150}{1296}$	$\frac{20}{1296}$	$\frac{1}{1296}$

1.5 Two standard results. In the working of Section 1.4, we have in fact used two standard results in probability. We have treated them as intuitively obvious, and shall continue to do so, without discussing whether or not we can prove them; but it is useful to state them explicitly here.

The first result is this:

If A and B are two *independent* events, that is to say, if the result of one does not affect the result of the other, then the probability of (A *and* B) is given by *multiplying* the individual probabilities, that is

$$p(A \ and \ B) = p(A) \times p(B).$$

This enables us to calculate the probability along a branch, that is, the probability of the exact sequence of results represented by the branch.

The second result is this:

If A and B are two *exclusive* events, that is to say, ones which cannot both occur simultaneously, the probability of (A *or* B) is given by *adding* the individual probabilities, that is

$$p(A \ or \ B) = p(A) + p(B).$$

This enables us to add together the probabilities of the branches which represent the outcomes with which we are concerned, and justifies the counting process involved in using Pascal's triangle.

1.6 The Binomial theorem. If we multiply out $(q+p)^3$ we shall get $q^3 + 3q^2p + 3qp^2 + p^3$. To obtain $(q+p)^4$, we take the product $(q+p)^3 \times (q+p)$, thus:

$$
\begin{array}{r}
q^3 \ + 3q^2p \ + \ 3qp^2 + p^3 \\
q \ + p \\
\hline
q^4 + 3q^3p + 3q^2p^2 + \ qp^3 \\
q^3p + 3q^2p^2 + 3qp^3 + p^4 \\
\hline
q^4 + 4q^3p + 6q^2p^2 + 4qp^3 + p^4
\end{array}
$$

The numbers 1, 4, 6, 4, 1 are the Pascal numbers (see above), and are often called the *Binomial Coefficients* when they arise in this way from the expansion of a *two-term* (binomial) expression $(q+p)$. The process of multiplication indicates another reason why each row can be deduced from the previous row by adding consecutive coefficients.

We can now write down the terms of $(q+p)^n$. The simplest way to obtain the *coefficients* is by extending Pascal's triangle far enough; the simplest check on the *powers* of q and p is that the sum of the indices must always be the index n under consideration.

When the theorem is applied to the situation represented by a tree diagram, $p+q$ is always equal to 1; hence p and q are the probabilities of two events of which it is certain that one or the other (but not both) will occur.

1.7 Notation for Pascal's triangle. The main difficulty about stating the Binomial theorem in the general case is in finding a suitable notation for the coefficients. We therefore label first the rows and then the individual numbers in Pascal's triangle as follows:

1	(row 0)	The numbers in row n of
1 1	(row 1)	Pascal's triangle are called
1 2 1	(row 2)	$\binom{n}{0}, \binom{n}{1}, \binom{n}{2}, \dots, \binom{n}{n}$
1 3 3 1	(row 3)	and their basic property is
1 4 6 4 1	(row 4)	
1 5 10 10 5 1	(row 5)	$\binom{n}{r} + \binom{n}{r-1} = \binom{n+1}{r}$.
1 6 15 20 15 6 1	(row 6)	

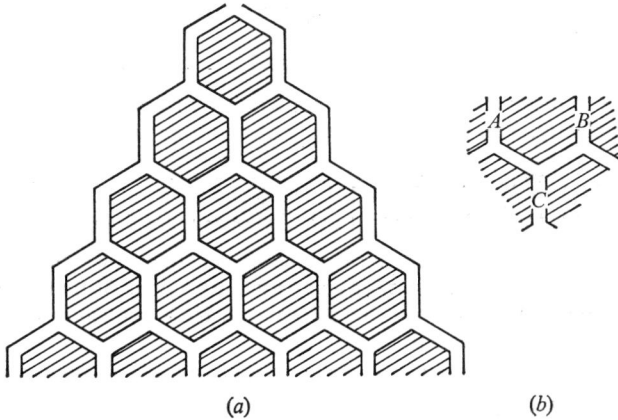

(a)   (b)

Fig. 7

The Galton Board, or Quincunx, (shown in Figure 7) helps to illustrate both this notation and the definition of further numbers by addition.

A ball is allowed to roll down a slope laid out as in Figure 7(a). If three points A, B and C are abstracted, as in Figure 7(b), and $N(X)$ is the number of distinct paths leading to X, moving always in a downward direction, it is clear that

$$N(A) + N(B) = N(C).$$

A little investigation will show that the number of paths leading to any such point is simply the appropriate number in the Pascal triangle; more precisely, if A is a point in the $(i+1)$th channel in the row containing m hexagons, then

$$N(A) = \binom{m}{i}.$$

If each path at a junction is equally likely to be taken by a ball, and a large number of balls are allowed to roll down the board, the numbers collected at the foot of each channel should then be roughly proportional to the Pascal triangle numbers, and the balls will form a pattern very like one of the histograms in Figure 5.

We may now identify the number $\binom{n}{i}$ with:

(a) the number of paths to the channel between the ith and $(i+1)$th hexagons in the row of the Galton Board containing n hexagons;

(b) the number of branches of a tree diagram which represents n independent trials, each with probability p of success and q of failure, with probability $q^{n-i}p^i$ 'along the branch';

(c) the coefficient of $q^{n-i}p^i$ in the expansion of $(q+p)^n$. We shall also show (Exercise A, Question 12) that it has the value

$$\frac{n(n-1)(n-2)...(n-i+1)}{1.2.3...i}.$$

Exercise A

If you have to make any assumptions not explicitly mentioned in the question, state them clearly. You need not multiply out your answers to Questions 3, 4, 5 and 7.

1. If one child in four is left-handed and I choose a random sample of five children, what is the probability that I shall choose more than two left-handers?

2. To allow postponements for inclement weather, four dates are chosen with a view to holding two rehearsals for an annual inspection. At that time of year, one day in four is too inclement for a rehearsal. How frequently will it be impossible to hold two rehearsals?

3. I reckon to be able to detect which of two different methods of brewing coffee is being used 95 % of the time. A test consists of six samples, all of which I must identify correctly. What are my chances of passing it?

4. A soccer goalkeeper stops shots with a probability of $\frac{3}{4}$. If there are 6 shots in the course of a game, what are the probabilities of 0, 1, 2, more than 2, goals being scored against him?

380

5. A firm buying nylon stockings in bulk is prepared to accept 10 % to be sold as 'seconds'. They inspect two random samples of ten pairs each, and find two pairs of 'seconds' in each. How likely is this (or a worse) result if only 10 % are in fact 'seconds'?

6. Write out the full expansion of $(q+p)^7$.

7. What are the values of p and q that you would use to predict the chances of throwing sixes at dice?
  What is the probability of scoring 48 in 8 rolls of a die?
  What is the probability of scoring 47 in 8 rolls of a die? (Do not simplify your answers.)

8. An industrial process turns out 5 % of defective valves, and fifty samples of five are inspected. What distribution of defectives would you expect among the samples?

9. One in ten of a population of apparently identical rabbits has a particular substance in its blood. What size of samples should a biologist require if he wants 95 % of the samples to contain at least one such rabbit?

10. Find the mean and the standard deviation of the number of successes given by the binomial probability pattern for the Doghouse missile.

11. (a) In n independent trials, where the probability of success in a single trial is p, the mean number of successes is np; and
  (b) the standard deviation of the number of successes is $\sqrt{(npq)}$, where $p+q = 1$.
  Confirm that these general results are true for the probability patterns given in Questions 1–5.

12. Show that, for $n = 7$, the following expressions give the correct Pascal triangle numbers:

$$\binom{n}{0} = 1; \qquad \binom{n}{1} = \frac{n}{1}; \qquad \binom{n}{2} = \frac{n(n-1)}{1.2};$$

$$\binom{n}{3} = \frac{n(n-1)(n-2)}{1.2.3}; \qquad \binom{n}{4} = \frac{n(n-1)(n-2)(n-3)}{1.2.3.4}.$$

Continue the sequence to confirm, in this case, the general formula

$$\binom{n}{i} = \frac{n(n-1)(n-2) \ldots (n-i+1)}{1.2.3 \ldots i}.$$

(Notice that the sum of corresponding factors, such as $(n-2)$ and 3, is always $(n+1)$.)
  Assuming that this formula gives

$$\binom{m}{r} \quad \text{and} \quad \binom{m}{r-1}$$

correctly, find the sum

$$\binom{m}{r} + \binom{m}{r-1}$$

and write down what the formula gives for $\binom{m+1}{r}$. Are these two numbers equal?

Hence show that if the formula gives the Pascal numbers $\binom{m}{i}$ correctly, for

$i = 0$ to $i = m$, it also gives the numbers $\binom{m+1}{i}$ correctly; and deduce that the formula always gives the same numbers as the addition rule by which the Pascal numbers were previously defined.

*13. The Binomial theorem asserts that

$$(q+p)^n = q^n + \binom{n}{1} q^{n-1}p + \binom{n}{2} q^{n-2}p^2 + \ldots + \binom{n}{i} q^{n-i}p^i + \ldots + p^n.$$

Is it true for the case $n = 7$?

Assuming that the theorem is true for the case $n = k$, form the product

$$(q+p)(q+p)^k,$$

and, assuming that $\qquad \binom{k}{i} + \binom{k}{i-1} = \binom{k+1}{i},$

show that the theorem is true for the case $n = k+1$.

Can you now deduce that the theorem is true for all positive integral values of n?

2. THE NORMAL PROBABILITY PATTERN

2.1 The Normal curve. In Figure 5, the histograms of binomial probability patterns with $m = 2, 3, 4, 5, 25$ were drawn, in such a way that the figures were comparable. To make this possible, three quantities had to be carefully controlled.

The *means* of the numbers of heads were all numerically different, because in each case they were equal to np (see Exercise A, Question 11). They could, however, be moved to the origin by a simple translation.

The *standard deviations* were given by the formula $s^2 = npq$, and they were represented in each case by the same length, $\frac{1}{2}$ unit.

The area under the histogram was kept at 1 sq unit. This represented a probability of 1 in each of these examples, and therefore made the vertical scales comparable.

The shapes of these diagrams are clearly approximating more and more closely to the bell-shaped Normal curve, drawn in Figure 5 on the same scale. It is interesting to fit a tracing of the curve over the histograms, to appreciate the exactness of the approximation.

The Normal curve may then be regarded as an idealization of all these histograms, and may be thought of as an abstraction from a situation where a large number of small decisions are made whose outcomes are equally likely.

Many naturally occurring measurements are the result of a large number of small decisions, and this probably explains why approximations to this pattern occur so frequently in measurements of natural phenomena. To take just two examples, the I.Q.s of children in a particular age range, and the weights of recruits to an army, are strikingly close to this pattern. In fact, it occurs so frequently that there is a real danger of supposing, quite unjustifiably, that unless there is a good reason to the contrary, the pattern that emerges will probably be Normal.

An interesting illustration of a Normal pattern, arising from an apparently unpromising background, is contained in Exercise B, Question 17. From a parent collection of numbers which can be represented by a rectangular histogram, (1–50), 5 numbers were drawn at random. The means of these 5 numbers, in a large number of such experiments, form a population which approximates to this Normal pattern. In fact, this kind of experiment always leads to a pattern which is nearly Normal, if the samples are large enough; and it may be taken as a simple mathematical parallel to what happens when we take measurements of natural phenomena.

The Normal curve has always an area of 1 beneath it. The mean, of course, lies on the axis of symmetry, and it is convenient to think of the curve as having the standard deviation as the unit of measurement on the horizontal axis. The following proportions are worth remembering and may be used freely:

68 % of the readings are within 1 standard deviation of the mean;
95 % of the readings are within 2 standard deviations of the mean;
$99\frac{1}{2}$ % of the readings are within 3 standard deviations of the mean.

The first of these results is the basis of the check in the previous chapter, that the Inter-Sextile Range approximates to 2 standard deviations. It is important to remember that this and the other properties listed here apply only to populations which are approximately Normal.

2.2   Cumulative frequency. If, therefore, we know the mean and the standard deviation of a population which is approximately Normal, we can in fact reconstruct a very fair approximation to the frequency table for the whole population. Statisticians prefer the *cumulative frequency* diagram for this purpose (see Figure 8); its construction was explained in earlier books.

Examples 3 and 4 illustrate the use that can be made of the diagram, if we express the deviation of each member of the population from the mean in units of one standard deviation.

2.3   Some examples.

Example 3. A Local Education Committee uses, for selection purposes, an intelligence test designed to give a mean of 100 points and a standard deviation of 15 points. 20000 children take this test and are to be graded in the following proportions:

the top 12 % for secondary grammar schools;
the next 28 % for bilateral schools;
the remaining 60 % for secondary modern schools.

(*a*) What will be the lowest expected I.Q. for a grammar school?
(*b*) What will be the lowest expected I.Q. for a bilateral school?

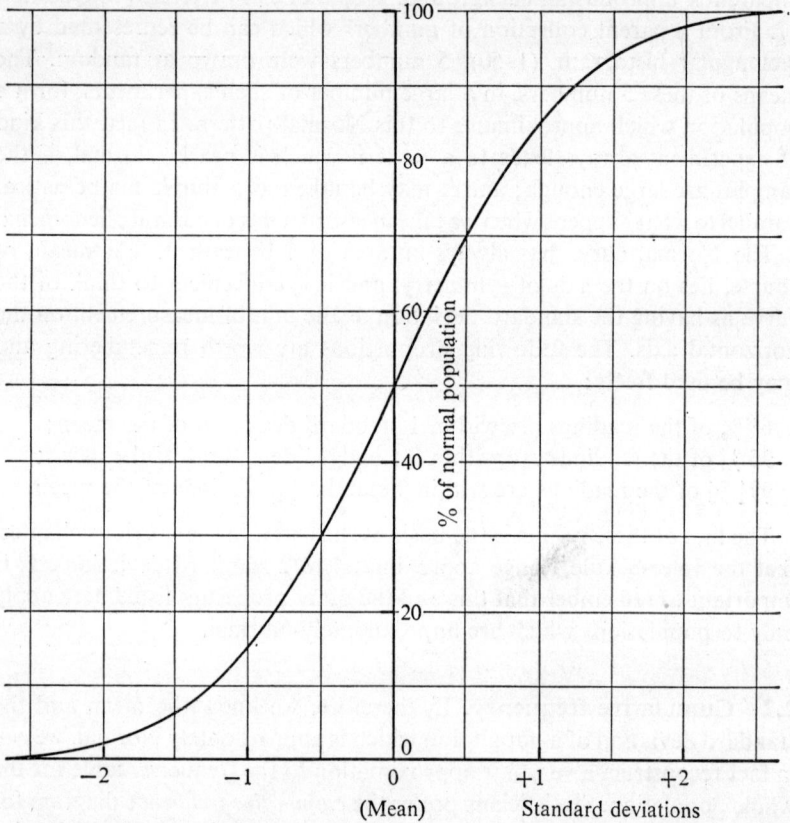

Fig. 8. Cumulative frequency diagram for a Normal population.

The committee requires that all children within 5 points of these lowest scores, above and below, shall be interviewed, because the accuracy of the test is liable to an error of 5 points either way.

(*c*) How many children will be interviewed?
(*d*) How inaccurate would the test in fact have to be before the same child could be 'borderline' for all three schools at the same time?

384

We assume the distribution to be Normal, with $m = 100$, and $s = 15$. We use the cumulative frequency diagram in Figure 8.

(a) Since 12 % of the children have I.Q.s above the dividing line between the grammar schools and the bilateral schools, 88 % of the children have I.Q.s below this. From Figure 8, we see that 88 % should lie below 1·2 standard deviations above the mean. Since each standard deviation represents 15 points, this means that the dividing line will lie 18 points above the mean. The lowest expected I.Q. for a grammar school is, therefore,

118 points.

(b) Similarly, 60 % of children will have I.Q.s below the second dividing line. From the graph, 60 % of children have I.Q.s up to 0·25 standard deviations above the mean. Each standard deviation represents 15 points, so that the lowest I.Q. that can be expected for a bilateral school will be 4 points above the mean, that is

104 points.

(c) The dividing marks are at 118 and 104 respectively, so that the borderlines extend from an I.Q. of 123 to an I.Q. of 113 in the first case, and from an I.Q. of 109 to an I.Q. of 99 in the second case. These four I.Q.s correspond respectively to 1·53, 0·77, 0·60 and −0·07 standard deviations above the mean. Now, Figure 8 shows that 94 % of children have I.Q.s less than 1·53 standard deviations above the mean. Similarly, 78 %, 73 % and 47 % of the children have I.Q.s below the other marks named.

Therefore (94−78) %, that is, 16 % of the children, lie in the upper borderline; and (73−47) %, that is, 26 %, of the children lie in the lower borderline. Thus, in all,

42 % (8400) of the children must be interviewed.

(d) The difference between the two dividing lines is 14 points. If, therefore, the test is liable to give a result wrong by more than 7 points, a child may well find itself in the borderline for all three schools at the same time.

Example 4. The chest measurements of Army recruits have mean 74·2 cm and standard deviation 3·8 cm. Uniform jackets are made in sizes: small; 68–72 cm; 72–76 cm; 76–80 cm; large. About how many per thousand will be expected to be in each class?

Assume the distribution to be approximately Normal, with $m = 74.2$ cm, $s = 3.8$ cm. The dividing line between the large size and the others is 80 cm, which is 5·8 cm above the mean; that is (5·8/3·8)s or 1·53s above the mean.

From the cumulative frequency diagram we see that 94% of the recruits will be less than 1·53s above m, so 6%, or 60 men per 1000 will be in the 'large' class.

The rest of the solution is shown in the table, set out thus:

Class division (cm)	Centimetres above mean (cm)	S.D. above mean	Cum. % below	% in class	Uniform class (cm)	Expected number
68	−6·2	−1·63s	5	5	Small	50
72	−2·2	−0·58s	28	23	34–36	230
76	+1·8	+0·47s	68	40	36–38	400
80	+5·8	+1·53s	94	26	38–40	260
—	—	—	100	6	Large	60

These results should be accurate to, say ± 10 men in each class.

2.4 Accuracy and validity. Accuracy of readings from the graph cannot be better than $\pm 1\%$, but tables can give the readings to $0·01\%$ and should be used when greater accuracy is needed and justified.

The heading of this section is The Normal Probability Pattern. The results of the examples are only valid if we assume that the population is Normal. Great care must be taken not to use this technique when the population is anything other than Normal. There are counter-examples in Exercise B to demonstrate this.

Exercise B

State clearly what assumptions you make in each question, and draw rough graphs to illustrate your answers.

1. If in this country the average I.Q. is 100 and the standard deviation 15, and the population is 51 million, how many people would you expect to find with I.Q.s:
  (a) over 125;   (b) under 80?

2. A commuter takes 43 minutes over his daily journey, with standard deviation 2·5 minutes. How long must he allow to make sure that he is punctual 98 % of the time?

3. Light bulbs have a mean life of 900 hours and a standard deviation of 80 hours. In a consignment of 500, how many would you expect to last
  (a) less than 1000 hours;   (b) more than 800 hours?

4. Morberry Jams Ltd. fill their $\frac{1}{2}$ kg jam jars with a machine which delivers 504·7 g on average with a standard deviation of 2·2 g. What percentage of jars will be under-weight? What weight of jam would they use in filling 2880 jars?

5. A fair coin is tossed 100 times. The mean number of heads is 50, with standard deviation 5. Between what limits (symmetrically placed about 50) will (a) 95 %, (b) 99 % of the results of such experiments lie?

386

6. The distribution of the weights of Ruritanian recruits is known to be Normal. The quartile weights are 130 and 160 kg. Estimate the mean and standard deviation and give the limits within which: (a) 95 % and (b) 99 % will lie.

7. In a standards competition, the high, medium and low standards for the 200 m race are 29 s, 30 s, 31 s. Of 156 entrants, 48 achieve high standards, 60 medium and 38 low. Show that these figures are consistent with a Normal distribution of ability (as measured by time taken), and suggest new standards so that 25 will achieve high standards and about 25 will fail to achieve low standards.

8. If you multiply the length of your foot (in cm) by $1\frac{1}{2}$, and subtract 25, the next whole number above the answer gives your shoe size. (For children's sizes subtract 11 only.)
  A retailer sells 2000 pairs of men's shoes and 4000 pairs of women's shoes a year. If the mean size of foot is 21 cm, with standard deviation 1 cm, for adult men, and 19 cm with standard deviation 1 cm, for adult women, how many of each size will he sell?

9. Imagine you want to obtain from your local shoe shop manager more accurate information for Question 8. What (non-technical) questions would you want to ask him?

10. For 2000 candidates in an examination, the mean mark was 54 %, the standard deviation 12 %, and the distribution approximately Normal. Estimate the number of candidates in each of the classes 1–10, 11–20, 21–30 and so on.

11. The girths of trees chosen at random from a plantation are as follows:

Under 30 cm	30–40 cm	40–50 cm	50–60 cm	Over 60 cm
3	17	117	56	7

Express these frequencies as percentages, and read off from Figure 8 how many standard deviations above or below the mean the divisions would lie if this were approximately Normal. Plot a graph of these values against the actual division marks, and hence estimate the mean and standard deviation. Check by direct calculation.

12. Think of a set of figures which you might reasonably expect to be Normally distributed, other than the ones in the text, and find some figures to see how, with the observed mean and standard deviation, the actual Normal population would be divided.

  In Questions 13–16, explain why the method described above would *not* give reasonable answers, and draw rough histograms to illustrate.

13. The mean age (both sexes) in the U.K. is 36, with standard deviation 22. Calculate the proportion over 80.

14. The mean of the numbers shown by a roulette wheel is 18, and the variance is 114. Calculate the proportion of '0's to be expected.

15. The mean score in a large number of throws of two dice is 7, with standard deviation approximately 2·4. Calculate the proportion of double sixes to be expected.

16. The number of misprints in a novel, per page, is as follows:

	0	1	2	3	4	5 or more
Frequency	94	95	47	16	4	1

Calculate the mean and standard deviation, and construct a Normal curve with this mean and standard deviation.

17. Divide $A = (1, 2, ..., 50)$ into ten subsets of five by a method which ensures that the division is random. Repeat the process four more times, and find the mean of each subset. Name this set of means B.

Calculate the m and s for A and for B.

Sketch histograms for the two sets on similar scales, marking m and $m \pm s$ on each. Comment on the shapes of the figures and the comparative values of the two means and the two standard deviations.

(*Hint*: try various sizes of class interval for set B; an interval of 2 or 3 will probably be most satisfactory.)

Miscellaneous Exercise on statistics and probability

(These questions are selected from S.M.P. Additional Mathematics papers set by the Oxford and Cambridge Examinations Board, and are reprinted by kind permission of the Board.)

1. Eight boys' marks in mathematics and in physics in an examination were as follows:

Maths	20	35	40	67	81	52	26	55
Physics	50	60	52	62	68	57	56	65

Calculate the mean and the standard deviation for the marks in each subject. Explain which you consider to be a better mark, 52 in mathematics, or 57 in physics.

Exhibit the results on graph paper with, for each boy, the mathematics mark as the x- and the physics mark as the y-coordinate. Comment on the pattern obtained. Another boy in the same group scored 53 in physics but missed the other examination. Estimate as fairly as possible the mark he would have been expected to score in mathematics and explain your procedure.

2. Plot accurately a cumulative frequency curve for the following data:

Distribution of marks in an examination

Mark range (%)	Under 10	10–29	30–49	50–59	60–69	Over 69
No. of candidates	1	9	26	39	14	11

Estimate from your curve the median mark. What is the significance of the steepest part of a cumulative frequency curve?

If the above figures correspond closely to the mark distribution in a large public examination in a certain year, and if the following year's distribution is notably different, state which you would regard as the sounder of the following possible procedures:

(i) to retain the pass and distinction marks at the same level; or

(ii) to award a pass and a distinction to the same proportion of candidates as before?

(Give reasons for your statement, making clear what assumptions you make.)

3. Explain what is meant by a coefficient of rank correlation. In a beauty competition two judges placed the contestants in the following orders:

(i) *C E D F A G I J B H*;   (ii) *F G D A I C H E J B.*

Find a coefficient of rank correlation between the two orders.

4. In a pocket game of cricket two hexagonal cross-sectioned metal pieces are rolled together. One piece is labelled with the scores 1, 2, 3, 4, 6, 4; and the other is labelled 1, 2, 3, 1, OUT, OUT. Find the distribution of the total scores from one throw of the two pieces (OUT rules out the score on the other piece, e.g. 2, OUT means a score of 0). State the probabilities of; (*a*) getting out; (*b*) scoring 5 or more; (*c*) scoring exactly 6.

Find the probability that a player survives two throws of the pieces and goes out on the third.

5. A die is in the shape of a regular tetrahedron. It has the numbers 1, 2, 3, 4 marked on its faces. After each throw the number which is not seen, i.e. the base-number, is recorded. A 'trial' consists of two throws of the die. Every possible result of a trial is a number-pair. Plot points to represent these, on a rough graph. Hence or otherwise determine the probability of each of the following:

(*a*) that the sum of the numbers does not exceed 5;
(*b*) that the sum is odd;
(*c*) that the difference does not exceed unity;
(*d*) that the numbers are the same.

Show that the standard deviation σ of the sum (score) is $\frac{1}{2}\sqrt{10}$ and that in a long series of trials less than half the scores would be expected to lie outside the range $5-\sigma$ to $5+\sigma$.

6. After fireworks have been stored in a certain climate it is found that on average a fifth of them will not light. A boy buys three. What is the probability that exactly two of them will light?

A man wishes to be 95 % sure (i.e. have a probability of 0·95) of having at least two which will light. Investigate whether four will be enough for him to buy.

7. John washes up after lunch 3 times a week and his sister Mary washes up the other 4 times. John's days are chosen at random each week. The probability that he will break one or more dishes during a washing is 0·1 and the probability that Mary will is 0·05. One day after lunch Dad, hearing a dish crash, said: 'Apparently this is John's day for doing the washing up.'

What is the probability that he was right?

8. (*a*) Three objects are placed in a row on a table. Four people in turn come in and carry out an interchange of *adjacent* objects. Follow through the various possibilities and hence find:

(i) the probability that a particular object is in the same place finally as initially;

(ii) the probability that all the objects are in the same positions finally as initially.

(Make clear what you assume about the actions of the people.)

(*b*) The five letters *p*, *q*, *r*, *s*, *t*, starting in this order, are deranged by a minimum number of adjacent interchanges:

(i) into the reverse order;

(ii) starting again as *pqrst* into the order *qpstr*. Calculate the number of moves in each case, and suggest how these numbers might be used to indicate 'degree of derangement'.

18

TIME: THE SUN, MOON AND PLANETS

1. THE SOLAR SYSTEM

The solar system consists principally of the central massive sun (diameter 1385000 kilometres) with nine planets moving around it in orbits, which, with one exception, lie practically in the same plane. The plane in which the planet earth orbits the sun is called the *plane of the ecliptic* and, as can be seen from the last column of the following table, the other planets (apart from Pluto) have orbital planes which are only slightly inclined to the ecliptic.

Further data about the planets is given in the table, which is arranged in order of their distances from the sun. These distances are given in astronomical units:

1 A.U. = earth's mean distance from sun = 1.495×10^8 kilometres.

Planet	Symbol	Mean distance from sun (A.U.)	Diameter (km)	Period of orbit	Inclination of orbit to ecliptic
Mercury	☿	0·387	5000	87·97 days	7°
Venus	♀	0·723	12000	224·7 days	3° 24′
Earth	⊗	1	12742	365·26 days	0°
Mars	♂	1·52	7000	686·95 days	1° 51′
		Minor planets. There are 2000 to 3000 of them of which Ceres (diameter 770 km) is the largest.			
Jupiter	♃	5·20	140000	11·86 years	1° 18′
Saturn	♄	9·54	120000	29·46 years	2° 29′
Uranus	♅	19·2	50000	84·02 years	0° 46′
Neptune	♆	30·1	55000	164·42 years	1° 46′
Pluto	♇	29·6–50	8–11000	248 years	17°

The planets, in their orbits, obey three simple laws discovered by Johannes Kepler (1571–1630) after prolonged study of data accumulated by the astronomer Tycho Brahe (1546–1601). Later it was shown by Newton that they could be derived by mathematical argument from general laws of motion set down in his *Principia*. A discussion of Kepler's laws follows their statement.

1.1   Kepler's laws of planetary motion.

I. Each planet moves in an elliptic path with the sun at one focus.

II. The line segment from the sun to a planet sweeps out equal areas in equal time intervals.

III. The squares of the periods of revolution of the planets are proportional to the cubes of the major axes of their paths.

1.2   Kepler's First Law: the ellipse.

An ellipse can be constructed as follows. A continuous piece of string is passed round two fixed pins *A*, *B* and held taut by the point of a pencil at *P*.

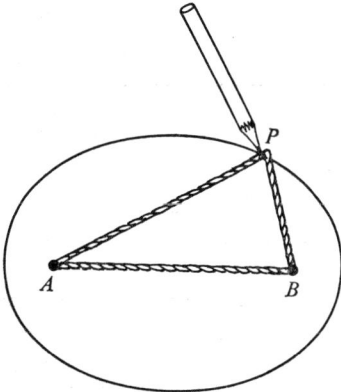

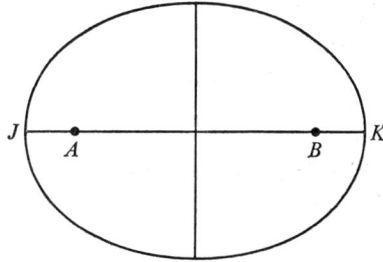

Fig. 1   Fig. 2

Keeping the string taut, the pencil, as it is moved round, traces out an *ellipse*.

The points *A* and *B* are the foci. *JK* is the *major axis* and the curve is symmetrical about this line (as it is also about the minor axis which bisects *JK* at right-angles).

By varying the length of string the ellipses can be made 'thinner' or 'fatter'. The number used to measure this property is known as the *eccentricity* (*e*);

$$e = \frac{AB}{JK} = \frac{AB}{AP+PB}.$$

The smaller the value of *e* the rounder the ellipse (see Figure 3). The eccentricity of the earth's orbit is $\frac{1}{60}$ (approximately) and it is small for all the major planets.

From this definition of an ellipse we can obtain its *polar equation*; if *P* is any point of the curve (see Figure 4), $AP = r$, angle $JAP = \theta$, then $AP+BP = c$, a constant $(= JK)$, and $AB = ec$, $BP = c-r$. Apply the cosine rule to the triangle *APB*: we obtain

$$(c-r)^2 = e^2c^2+r^2+2ecr \cos \theta,$$

391

since $\qquad\qquad \cos PAB = \cos(\pi-\theta) = -\cos\theta.$

Expanding each side of this equation and rearranging, we find that

$$c^2(1-e^2) = 2cr(1+e\cos\theta)$$
$$\Leftrightarrow r(1+e\cos\theta) = c(1-e^2)/2,$$

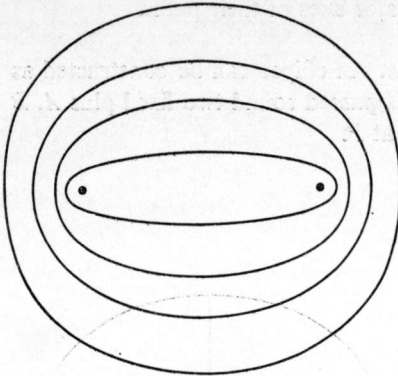

Fig. 3

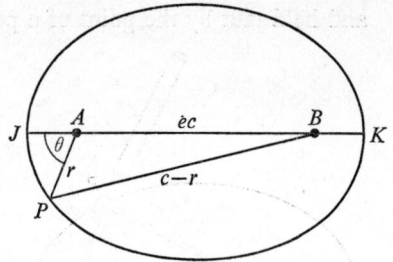

Fig. 4

which is a constant, usually denoted by l. This is the value or r when $\theta = \tfrac{1}{2}\pi$, and we finally obtain for the polar equation

$$l/r = 1 + e\cos\theta.$$

1.3 Kepler's Second Law: areal velocity. The planet in its positions of greatest and least distance from the sun (which occur at the ends of the major axis) is said to be in *aphelion* and *perihelion*, respectively. (The corresponding positions for the moon, or an artificial satellite, in its orbit

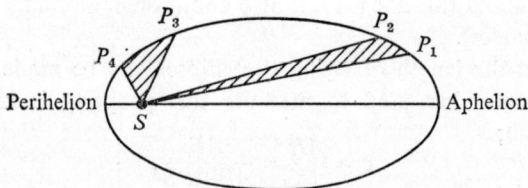

Fig. 5

about the earth, or of the sun in its orbit as seen from the earth, are apogee and perigee.) In Figure 5, P_1P_2 and P_3P_4 represent the movements of the planet in its orbit on two occasions over equal intervals of time. By Kepler's Second Law the shaded areas are equal. It follows that the speed of the planet varies continuously during its orbit about the sun, being least at aphelion and greatest at perihelion.

392

The exact law for the angular velocity of the planet about S is easily found. For if the angle $P_1 SP_2$ is small ($= \delta\theta$), the area of the sector $P_1 SP_2$ is approximately $\frac{1}{2}r^2.\delta\theta$ (see p. 241), and if this is swept out in a small time δt, the rate of sweeping out of area is the limit of the ratio of $\frac{1}{2}r^2\delta\theta$ to δt, i.e. $\frac{1}{2}r^2\, d\theta/dt$. Since this is a constant, the angular velocity $d\theta/dt$ is given by the equation

$$r^2\, d\theta/dt = h, \quad \text{where } h \text{ is a constant.}$$

This equation, and the equation of the orbit in the last section, describe the motion completely; Newton showed that they are immediate consequences of his law of universal gravitation, the force of attraction between the sun and the planet being along PS and inversely proportional to r^2.

1.4 Kepler's Third Law may be illustrated with reference to Mercury, Venus, and the earth. The law states that $T^2 \propto D^3$, where T is the orbital period and D the length of the major axis. This can be written as

$$T^2 = k.d^3,$$

where $d = \frac{1}{2}D$, the mean distance from the sun, and k is a constant for the solar system.

Now using the data from the table on p. 390, we obtain as values of T^2/d^3 for Mercury, Venus and the earth, the fractions

$$\frac{(87\cdot97)^2}{(0\cdot387)^3}, \quad \frac{(224\cdot7)^2}{(0\cdot723)^3}, \quad \frac{(365\cdot26)^2}{1^3},$$

where we have taken the orbital period in days and the mean distances in astronomical units. These fractions, computed to 4 significant figures, are 133600, 133700 and 133500; in good agreement with the law that $T^2/d^3 = $ constant. The slight variations in these three estimates of k are in part due to the fact that the law is only valid on the assumption that the mass of the planet is negligible in comparison with the mass of the sun.

Note. The moon and artificial earth satellites also obey Kepler's laws if one reads 'earth' for 'sun'. A new proportionality constant will of course be necessary in the corresponding third law.

Exercise A

1. Draw, in the same figure, ellipses of eccentricity $\frac{1}{2}$ and $\frac{1}{5}$ with their foci 4 cm apart.

2. Draw (to a stated scale) the orbit of the earth about the sun (eccentricity $\frac{1}{60}$).

3. Check that the orbits of Mars, Jupiter and Saturn, mutually satisfy Keplers' Third Law, approximately.

4. A hovering satellite has a fixed height of about 35200 kilometres. If another artificial satellite, having a period of 3 hours, has a least height of 480 kilometres above the earth's surface, what is its greatest height? (The radius of the earth is about 6400 kilometres.)

5. Evaluate the constant in Kepler's Third Law for the moon in relation to the earth, using the approximate value of 758000 km for the length of the major axis of the orbit. Hence calculate the period of an artificial satellite whose greatest and least heights above the earth's surface are 32000 and 320 kilometres

6. Sketch the curve given by the polar equation

$$r(3+\cos\theta) = 12.$$

What is the length of its major axis? What is its eccentricity?

7. Sketch the curves given by the equation

$$r(1+e\cos\theta) = l$$

(a) when $e = 1$; (b) when $e = 2$.

(These curves are also possible orbits, but since they extend to infinity, they can only apply to bodies entering the solar system from outside, or which are being lost to it.)

8. Find the ratio of:
(a) the angular velocities;
(b) the actual speeds, of the earth in its orbit at perihelion and aphelion, given the equation of its orbit (in A.U.) as $r(60+\cos\theta) = 60$.
[Use $r^2\,d\theta/dt = h$.]

9. [For discussion.] The Astronomical Unit of distance is based upon the dimensions of the earth's orbit; should a less geocentric choice have been made?

1.5 Scale model of solar system. If the solar system is represented on a scale where the sun becomes a sphere of diameter 2·5 cm the distances of the planets from the sun are very roughly:

Mercury 1 m	Venus 2 m	Earth 3 m
Mars 4½ m	Jupiter 15 m	Saturn 30 m
Uranus 60 m	Neptune 90 m	Pluto 120 m

On the same scale the nearest star is about 530 kilometres from the sun! It can be appreciated from this illustration that the whole compass of the solar system is minute in comparison with stellar distances. This relationship between the solar system and the surrounding stars can be illustrated again in the following way.

Picture a model of the sun in the middle of a moderate-sized room of a house and let the scale be such that the nearest stars are as far away as the walls, then the diameter of the earth's orbit about the sun is of the order of 1/400th of a centimetre. The configuration of the stars will seem very much the same from whichever part of the earth's orbit they are observed. There is of course a very small shift of the nearer stars compared with the more distant ones due to the earth's movement round the sun throughout the year, and in the few cases where it is measurable it gives a means of estimating the distances to those stars. The *parallax* of a star is in fact half the maximum angular shift caused in this way.

1.6 The earth in its orbit. The earth moves round the sun in an elliptic path which is nearly circular; the sun being at one focus. On about 3 January, when the earth is at perihelion, the sun is 147 000 000 km away, and this distance increases to 152 000 000 km at aphelion on about 5 July.

The polar axis of the earth is not at right-angles to the ecliptic (that is, the plane of the earth's orbit) but is inclined from this direction at about 23° 27′. It must be remembered that, as well as its annual revolution about

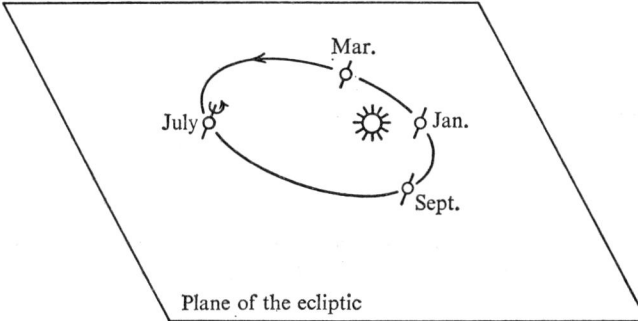

Fig. 6

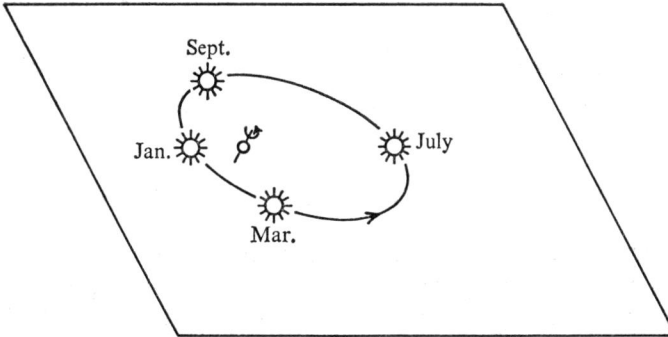

Fig. 7

the sun, the earth has a daily rotation about its polar axis—the rotation and the revolution being in the same sense as indicated by the arrows. Owing to this double movement of the earth, the relative movements of the sun, the moon, the planets and the stars, as observed by us from an apparently fixed earth, are somewhat complicated.

Figure 6 shows the movement of the earth relative to the sun in a space whose fixed directions are those of the distant stars. Remembering the smallness of the solar system in comparison with stellar distances we can derive the motion of the sun relative to the earth in relation to what are effectively the same fixed directions; Figure 7 shows this relative situation.

It will be seen that at any time the vector from earth to sun is the same in both figures. It follows that the sun, relative to the earth, moves in an ellipse equivalent to that of the earth's path relative to the sun. The ellipse is traversed in the same sense but now the earth occupies the other focus.

1.7   The seasons. The last figure is drawn with the earth's north pole 'above' the plane of the ecliptic. At a particular instant of a day in July, the place on the earth which has the sun directly overhead (that is, the place where the join of the centres of the earth and sun cuts the earth's surface) is in the northern hemisphere. As the earth rotates once, the locus of such subsolar places is a parallel of latitude (neglecting the small movement of the sun in its orbit during this day). At this time of the year it is summer in the northern hemisphere and winter in the southern hemisphere. On a day in March and September the locus of subsolar places is the equator, while in January it is a southerly parallel of latitude.

The most northerly such parallel of latitude is called the tropic of Cancer (23° 27′ N) and the most southerly the tropic of Capricorn (23° 27′ S). The positions of the sun in its orbit at these two extremes, on about 21 June and 22 December, are the so-called summer and winter solstices (summer or winter depending on the particular hemisphere of the observer), and they correspond to the longest and shortest days—at least for latitudes between the Arctic Circle (66° 33′ N) and Antarctic Circle (66° 33′ S). Within the Arctic and Antarctic Circles, the sun is above the horizon all day at mid-summer for a number of days depending upon the latitude.

The occasions when the sun is directly above the equator are known as the spring and autumn equinoxes (on about 21 March and 22 September), when, as the name implies, day and night are of equal duration.

1.8   The zodiac. When considering the apparent movements of heavenly bodies as seen by an observer on the earth we must remember three things: (i) the opacity of the earth allows us to see only one half of the sky at any instant; (ii) when the sun is in that hemisphere, the stars are invisible; (iii) the apparent rotary motion of the whole sky from east to west is due to the earth's rotation on its axis in the opposite direction.

Therefore to simplify the situation let us suppose that the earth stops spinning, that the luminosity of the sun is weakened to about that of the moon, and that the earth is reduced to a very small size so that an observer could have a more or less all round view of the sky. He would see a fixed spherical background of stars with the sun slowly moving eastwards over it, following a fixed great circle route, traversed annually. This track is called the celestial ecliptic. The belt of the heavens lying within some 8° either side of the ecliptic is the zodiac, and within this belt are to be found the moon and the visible planets. The moon also would be seen to move

396

eastwards past the stars, taking a month for each circuit; whereas the planets would show a more irregular although mainly easterly movement.

Re-introducing the easterly rotation of the earth is relatively equivalent to superimposing an equal westerly rotation on the movements so far described. In particular, the resulting apparent motions of the sun and the moon are westwards but the moon will be seen to move more slowly than the sun in this direction because it actually has the faster easterly movement as seen from a non-rotating earth.

The zodiac is divided into twelve equal parts—the 'signs'—and, in the order in which the sun passes through them, they are:

I. Aries (ram)	♈	II. Taurus (bull)	♉
III. Gemini (twins)	♊	IV. Cancer (crab)	♋
V. Leo (lion)	♌·	VI. Virgo (virgin)	♍
VII. Libra (scales)	♎	VIII. Scorpio (scorpion)	♏
IX. Sagittarius (archer)	♐	X. Capricornus (goat horn)	♑
XI. Aquarius (water carrier)	♒	XII. Pisces (fishes)	♓

When first named the signs contained the constellations of the same names but, as we shall see later in Section 2.4, they are now 'one removed', for example, the constellation of Aries is now in the sign Taurus.

2. THE CELESTIAL SPHERE

2.1   The celestial sphere. In Section 1.5 it was emphasized that the whole of the solar system is quite minute when considered in relation to stellar distances and hence the configuration of the stars is appreciably the same viewed from any part of the solar system.

It has previously been found helpful to suppose that the earth was not spinning, that the sun was not very bright and that the earth was shrunk to a very small size so that an observer there could get an all round view of the sky. To such an observer the background of stars would appear to be printed on a huge fixed sphere of which he is the centre; this sphere we call the celestial sphere.

More formally we can define the celestial sphere as a sphere whose centre is the centre of the earth and whose radius is large compared with stellar distances. If we join the centre to the centre of a star, the point where the line produced meets the celestial sphere is called the position of the star on the celestial sphere, and this process is described as projecting the star on to the celestial sphere. We can similarly project any other heavenly body, such as the sun or moon or a planet, by joining the centre of the earth to its centre and letting the line cut the imaginary celestial sphere.

We can think of the celestial sphere as being essentially a fixed sphere because, although the stars do have proper motions of their own through space, their general configuration alters extremely slowly—the constellations remain very much the same shape over thousands of years. The sun, moon and planets, however, which are relatively close to us change their directions from us at an appreciable rate, so that their projected positions on the celestial sphere are not fixed.

Frequently when the context permits, we shall for convenience speak of the projection of the sun on the celestial sphere as the 'sun' and similarly for other heavenly bodies. The projection of the poles of the earth and of the equator give us the celestial poles and the celestial equator and so on. Here also the word celestial will be omitted at times when it is safe to do so.

Within the celestial sphere, at its centre, is the earth spinning eastwards daily on its polar axis. On the scale of the celestial sphere the earth should be represented as a mere point, but at times as in Figure 8, it will be useful to exaggerate its size.

Figures of the celestial sphere are really pictures of a model seen from the 'outside'. This will have to be borne in mind when interpreting such figures in terms of what is actually seen in the sky from 'inside'.

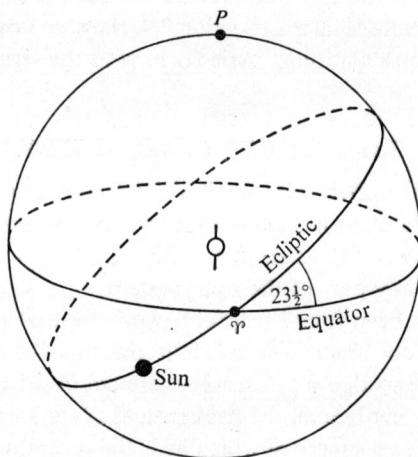

Fig. 8

2.2 The first point of Aries. Figure 8 shows the celestial north pole, celestial equator and celestial ecliptic. The celestial position of the sun travels round the celestial ecliptic once a year, and the earth rotates on its axis once a day.

Now (dropping the word celestial) where the sun in its annual journey crosses the equator from south to north is the position known as the first point of Aries, and is represented by the sign ♈.

As the earth's axis is tilted away from the perpendicular to the ecliptic by about $23\frac{1}{2}°$, the two great-circles of the equator and ecliptic cut at this angle.

398

The sun is at Aries at the spring (or vernal) equinox about 21 March, reaches its furthest north position about 21 June, is at Libra at the autumn equinox about 23 September and reaches its furthest south position about 21 December. The furthest north and furthest south positions are said to occur at the solstices.

2.3 Position on the celestial sphere: right ascension and declination. Position on the earth is defined by longitude and latitude which are measured respectively from the meridian of Greenwich and from the equator. Positions of heavenly bodies on the celestial sphere are defined analogously, using the celestial equator and the meridian through ♈ as

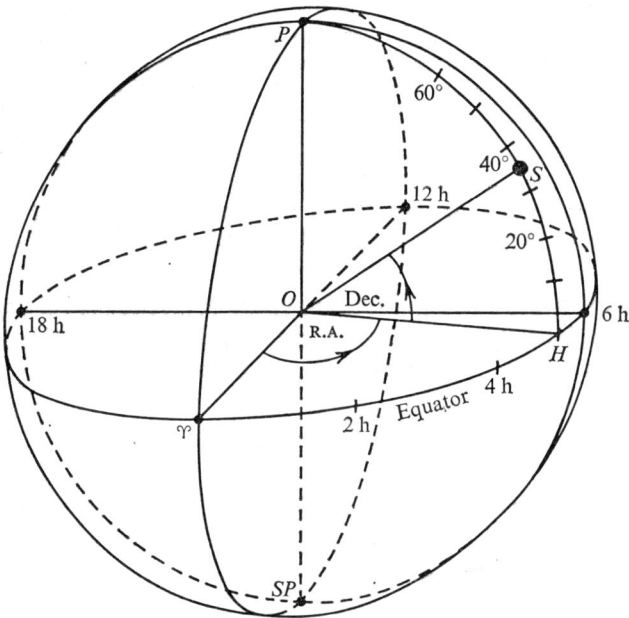

Fig. 9. Celestial sphere.

axes. Angular distance above or below the equator is termed declination (Dec.). Angular distance between the meridian through ♈ and the meridian of the heavenly body is termed right ascension (R.A.).

Declination is measured in the same way as latitude.

Right ascension is measured from ♈ to the east in hours, minutes and seconds on the scale 24 h = 360°, 1 h = 15°, 4 min = 1°.

In the figure, for example

R.A. of S is 5 h 30 min,

Dec. of S is 35° N.

Exercise B

1. Estimate roughly the right ascension and declination of the sun in May.

2. Estimate roughly the right ascension and declination of (i) a full moon, (ii) a moon 'one week old', in the third week in May.

3. On a figure of the celestial sphere show the relative positions of the sun and Capella (R.A. 05.14, Dec. 46° N) at the beginning of April. Would Capella be visible to an observer near the equator as a morning star, an evening star or not visible to the naked eye on a clear night?

4. At about what time of the year will Vega (R.A. 18.35, Dec. 39° N) be on an observer's meridian at midnight?

5. Name a month of the year when it would be a waste of time for anyone to look for Aldebaran (R.A. 04.33, Dec. $16\frac{1}{2}$° N) on a clear night. What is the maximum altitude of this star for an observer in latitude 52° N? (Altitude is the angular height of a heavenly body above the horizon, and is greatest when the heavenly body is nearest to the observer's zenith.)

6. The 'pointer stars' are two stars of the Great Bear constellation which are approximately in line with the pole star. Does this line appear to move clockwise or anticlockwise in the sky? If on a clock face this line reads 3 o'clock at midnight tonight, at what time will it read 4 o'clock after the lapse of a month?

2.4 Precession of the equinox. The centre of the head (T) of a spinning top, whose point (O) is not allowed to move, will trace out a circular path (LM), when the motion is steady. If it were not spinning the weight of the

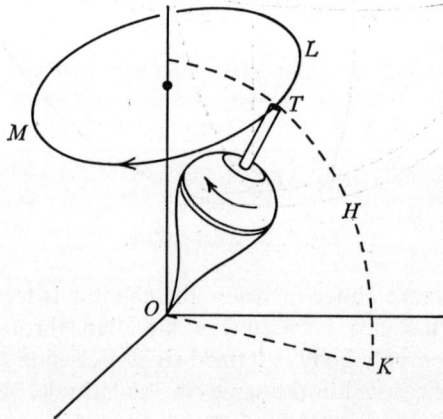

Fig. 10

top would cause the head to fall along the path THK, but when it is spinning the gyroscopic effect is to cause the head to move at right angles to this path and to travel horizontally round the circle LM in the direction of the arrow (when the top's spin is in the direction shown by the other arrow). This motion is known as 'precession'.

If the earth were perfectly spherical the attraction exerted on it by the moon or sun would be equivalent to a force through its centre. But the earth is not spherical—it bulges out at the equator—so we may think of it as a sphere with an additional layer (shown shaded in Figure 11). The part A being nearer the moon (or sun) is attracted more strongly than the opposite part B, so that the moon's (or sun's) attraction provides a torque which tends to align the earth's axis in a direction perpendicular to the

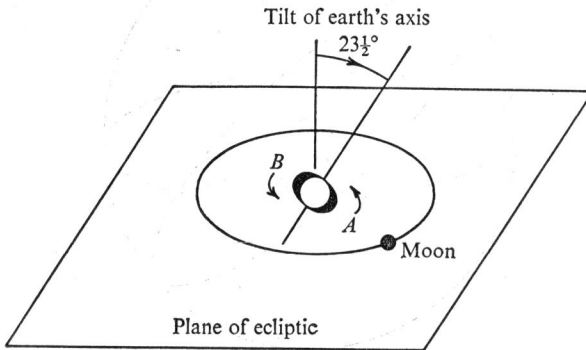

Fig. 11

orbital planes which we take to be the plane of the ecliptic. Because of the earth's axial spin this tendency is, as it were, diverted into a precessional motion of the axis similar to that for the top. The tendency of the earth's axis to right itself is contrary to that of the axis of the top but as the spins are also in opposite directions, the direction of precession is the same for both.

The torque produced by the moon or sun varies for different positions in their orbits; it will be greatest when the declination of the moon or sun is greatest and zero when the declination is zero, but (except when it is zero) it always tends to right the earth's axis. The maximum torque produced by the moon is about twice that of the sun. If we consider only the average long-term effect, the result is that the celestial pole, P, moves slowly in a small circle in the direction indicated in Figure 12, taking about 26000 years to complete the circuit. The variation about this average path gives a small sinuosity to this motion and is known as 'nutation'.

It happens, in the present era, that the celestial pole almost coincides with a star which we call 'Polaris' (α Ursae Minoris). It will move away from this (see Figure 13) and in about A.D. 14000 will point roughly towards the bright star Vega, whereas in about 3000 B.C., α-Draconis would have served as pole star. The centre A of the circular path of the

401

pole in the sky is called the 'pole of the ecliptic' and is approximately in the middle of the constellation Draco.

Now it can be seen from Figure 12 that as the pole P precesses, so the first point of Aries ♈, at the intersection of the equator and ecliptic, will move round the ecliptic in the same westerly direction. This movement of Aries is the so-called 'precession of the equinox'.

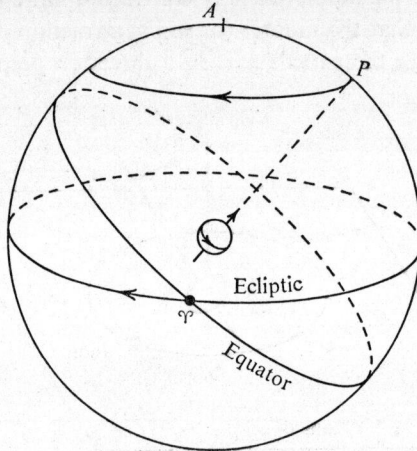

Fig. 12

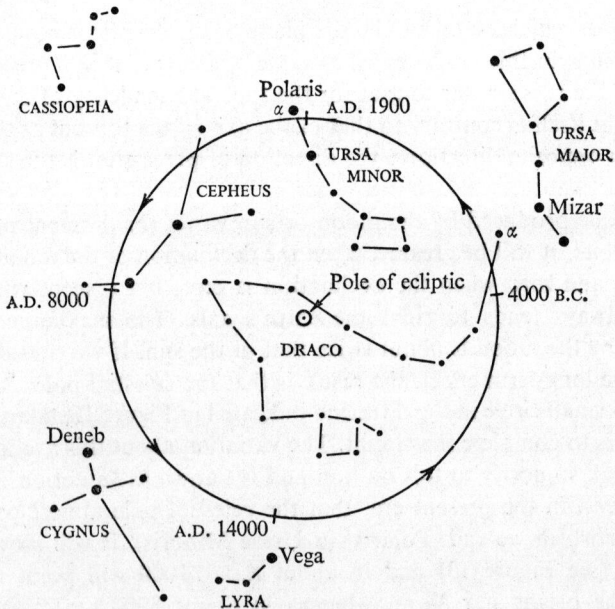

Fig. 13

The first point of Aries, which was originally so named because it was then in the constellation of Aries, has since precessed into the neighbouring constellation of Pisces. The equally spaced signs of the zodiac bear the names of the constellations in which they originally lay, and as the first sign Aries has shifted into Pisces it has caused all the zodiac signs to be shunted into the constellations one to the west—an unwelcome move for the astrologers.

The fact of precession, but not its cause, was known in ancient times. The Greek astronomer Hipparchus in 130 B.C., by comparing his observations with some recorded 150 years earlier, showed that the star Spica had increased its distance from the first point of Aries by about 2°, and that there was evidence that all the stars, relative to Aries, were moving parallel to the ecliptic. He attributed this correctly to a slow movement of the equator, whose inclination to the ecliptic remained constant.

Exercise C

1. Estimate the angular difference between the celestial pole and Polaris in 600 B.C. Is it likely that sailors of that time would have used Polaris as an aid to navigation?

2. Chaldean astronomers observed stars which are now too far south to be observed from their country. Explain this fact.

3. TIME

3.1 Sidereal and solar time. Time, in an everyday sense, is closely related to the position of the sun in the observer's sky. At the instant depicted in Figure 14(a), the sun (S) and a star (T) both lie on the celestial

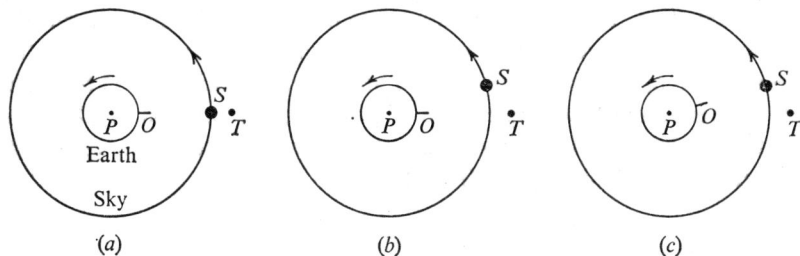

Fig. 14

meridian of an observer (O), and P is the north pole of the earth. For the observer, the sun and star would be either overhead or due north or south of him and are then said to be 'in transit across his meridian'.

Figure 14(b) shows the situation when the observer is next in line with the star, that is after a 'sidereal' day. By this time, the sun in its yearly motion round the sky will have shifted through about 1° (360° in 365 days). Therefore in approximately a further 4 minutes the observer will come into line with the sun as in Figure 14(c). By that time the earth will have made in all one complete rotation relative to the sun, in an interval known as a 'solar' day.

As seen by an observer, who instinctively reckons the earth to be still, the stars appear to rotate to the westward and the sun also appears to move in that direction but a little more slowly. One would perhaps be more aware of their relative movements if the sun and the stars were visible at the same time.

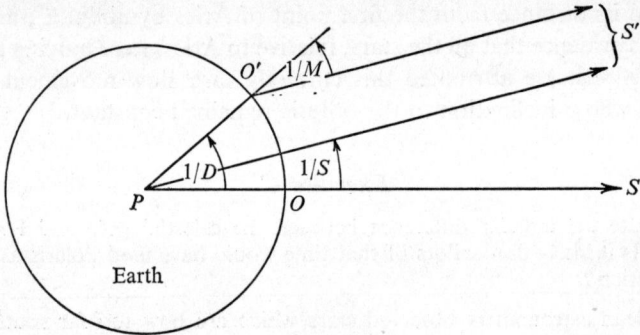

Fig. 15

We may now consider this more precisely. Let S be the time the earth takes to complete one revolution of its orbit relative to the distant stars, i.e. to pass from its position in the line ST to its next position in this line. This is the *sidereal year* of 365·25636 days. Let D be the length of the sidereal day—the time of one revolution of the earth on its axis, relative to the distant stars. Then in one unit of time the vector **PS** will have turned through $1/S$ of a revolution, the vector **PO** will have turned through $1/D$ of a revolution, and the vector **OS** will have turned relative to **PO** through the difference of these angles (see Figure 15), i.e. through $(1/D-1/S)$ revolutions. Now a solar day is the time taken by the vector **OS** to complete one revolution; if this time is M we have

$$\frac{1}{D} - \frac{1}{S} = \frac{1}{M},$$

which is called the *equation of synodic motion*. The second is defined as 1/86400 of the mean value of M; D is a little shorter than M, since $D = SM/(S+M)$.

3.2 The mean solar day. An observatory clock, driving a telescope for star observation, keeps in step with the movement of the stars. As this movement is entirely due to the earth's steady rotation† on its axis, it is quite regular; such a clock is keeping sidereal time. This time rate is, for obvious reasons, of no use for normal every-day purposes; normal clocks must be regulated to keep pace with the alternations of day and night imposed by the sun. Unfortunately, the sun's apparent yearly motion across the star background of the sky is uneven, so that a solar day, measured between successive transits of the sun across a meridian, is of varying duration. For our time-keeping the 24 h of our clocks is the average length of a solar day and is called a 'mean solar day'.

There are two principal reasons for the sun's imperfect timekeeping: (a) the variable speed of the earth in its orbit about the sun, which relative to us is equivalent to the variable speed of the sun in its apparent orbit about the earth (see Section 1.3); and (b) the inclination of the ecliptic to the equator. The smoothing out of the sun's uneven motion caused in these two ways will now be considered in more detail.

3.3 The earth's variable angular velocity. We refer back to Figure 7, which shows the sun's orbit relative to the earth. This can be derived from the figure of the earth's orbit relative to the sun by a half-turn about the centre of the ellipse; if the vector **ES** is of length r, and $d\theta/dt$ is its angular velocity, we still have (Section 1.3)

$$r^2 \, d\theta/dt = h,$$

and
$$r(1 + e \cos \theta) = l.$$

From these equations we obtain

$$d\theta/dt = h/r^2$$
$$= h(1 + e \cos \theta)^2/l^2$$
$$= k(1 + 2e \cos \theta),$$

where we have put $h/l^2 = k$ and neglected the term in e^2, since e is small. As a first approximation we have $d\theta/dt = k$, so that $\theta = kt$; inserting this on the right-hand side, we obtain

$$d\theta/dt = k(1 + 2e \cos kt),$$

so that, integrating, $\qquad \theta = kt + 2e \sin kt.$ \hfill (1)

The zero of time, $t = 0$, in this equation, is the moment when $\cos \theta = 1$, i.e. the moment of the earth's perihelion passage on 3 January. The time between successive passages of perihelion, which is almost but not quite equal to the sidereal year S, is $2\pi/k$, this value of t giving $\theta = 2\pi$ in this equation.

---

† Sidereal and mean solar units of time, which depend on the rotation of the earth, are not in fact constants as the earth does not rotate at a constant rate (but the variations are extremely small).

3.4   The inclination of the ecliptic. The angle θ in equation (1) of the last section is of course the angle swept out by the vector **ES** in the plane of the earth's orbit. On a diagram of the celestial sphere (Figure 16), θ is the angular distance travelled by the sun along the ecliptic. The apparent time of day, however, is governed by the moment of the passage of the sun across the meridian, and this depends on the angular distance travelled by the sun's celestial meridian PS along the celestial equator, and is conveniently measured from the first point of Aries. Let us call this angle ϕ,

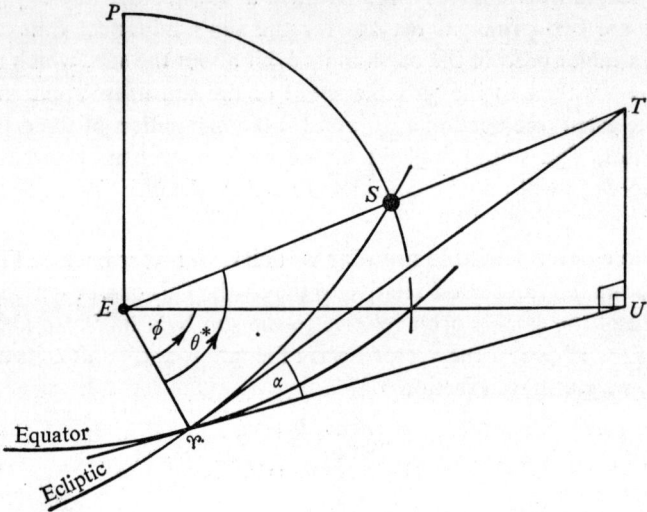

Fig. 16

and let us also call the angle $\Upsilon ES\ \theta^*$, which merely differs from θ by a constant. Now, in Figure 16, if $E\Upsilon$ is the unit of length, then $\Upsilon T = \tan\theta^*$, from the right-angled triangle $E\Upsilon T$, and $\Upsilon U = \tan\phi$, from the right-angled triangle $E\Upsilon U$. But the angle $U\Upsilon T =$ the angle of inclination of the ecliptic to the equator = the angle of tilt of the earth's axis = $23° 27'$. Hence

$$\tan\phi = \tan\theta^*.\cos\alpha, \quad \text{where} \quad \alpha = 23° 27'.$$

From this we have

$$\tan(\theta^*-\phi) = \frac{\tan\theta^* - \tan\phi}{1 + \tan\theta^* \tan\phi}$$

$$= \frac{\tan\theta^* (1 - \cos\alpha)}{1 + \tan^2\theta^* \cos\alpha}$$

$$\simeq \frac{2\tan\theta^* \sin^2\left(\tfrac{1}{2}\alpha\right)}{1 + \tan^2\theta^*},$$

replacing $\cos\alpha$ (0·919) in the denominator by 1,

$$\simeq \sin 2\theta^* \sin^2\left(\tfrac{1}{2}\alpha\right).$$

(We have used here several trigonometrical identities, which have not all been established in Chapter 14. The reader whose knowledge of trigonometry is limited must be prepared to take them on trust.)

Now $\sin^2(\tfrac{1}{2}\alpha)$ is very small (approximately 0·041), and the tangents of small angles are very nearly equal to their radian measures. This means that we can write

$$\theta* - \phi = 0\cdot041 \sin 2\theta*, \text{ approximately,}$$

or
$$\phi = \theta* - 0\cdot041 \sin 2\theta*. \tag{2}$$

Finally, we put the results (1) and (2) of this section and the last section together. If t is the time elapsed since the earth's perihelion passage, and $t*$ is the time elapsed since the vernal equinox when the sun passed the first point of Aries, then the angular distance travelled by the sun between these two events is

$$\theta - \theta* = k\,(t - t*) + 2e \sin k(t - t*), \quad \text{a constant.}$$

This means that approximately

$$\phi = kt* - 2e \sin k(t - t*) + 2e \sin kt - 0\cdot04 \sin 2kt*, \tag{3}$$

where we have replaced $\theta*$ in the last small term by $kt*$ which is nearly equal to it.

3.5   The mean sun. If we imagine a fictitious sun which traverses the celestial equator once a year at a steady rate, in such a way that on the average it is as often behind the true sun as ahead of it, its right ascension would be $kt* - 2e \sin k(t - t*)$ in equation (3) above. Such a sun is called the *Mean Sun*, and time as measured by it is *Mean Solar Time*. Since the true sun's right ascension is given by ϕ in equation (3), the difference in angle along the celestial equator between the positions of the true sun and the mean sun is given by

$$2e \sin kt - 0\cdot04 \sin 2kt* \quad \text{approximately,}$$

the true sun being ahead of the mean sun by this amount. Since the earth in its daily rotation on its axis is 'chasing' the sun, this means that the true sun comes to the meridian *later* than the mean sun by approximately

$$\frac{1}{2\pi}(2e \sin kt - 0\cdot04 \sin 2kt*) \times 24 \text{ hours.} \tag{4}$$

This quantity is called the *Equation of Time*, and it is the difference, in time units, (R.A. of True Sun − R.A. of Mean Sun). The period of $\sin kt$ is one year, and its zero is at the time of perihelion on 3 January; the period of $\sin 2kt*$ is therefore half a year, and its zero is at the vernal equinox on 21 March. The equation of time is plotted in Figure 17. To a fair degree of approximation it is given by

$$(7\tfrac{3}{4} \sin kt - 10 \sin 2kt*) \text{ minutes.}$$

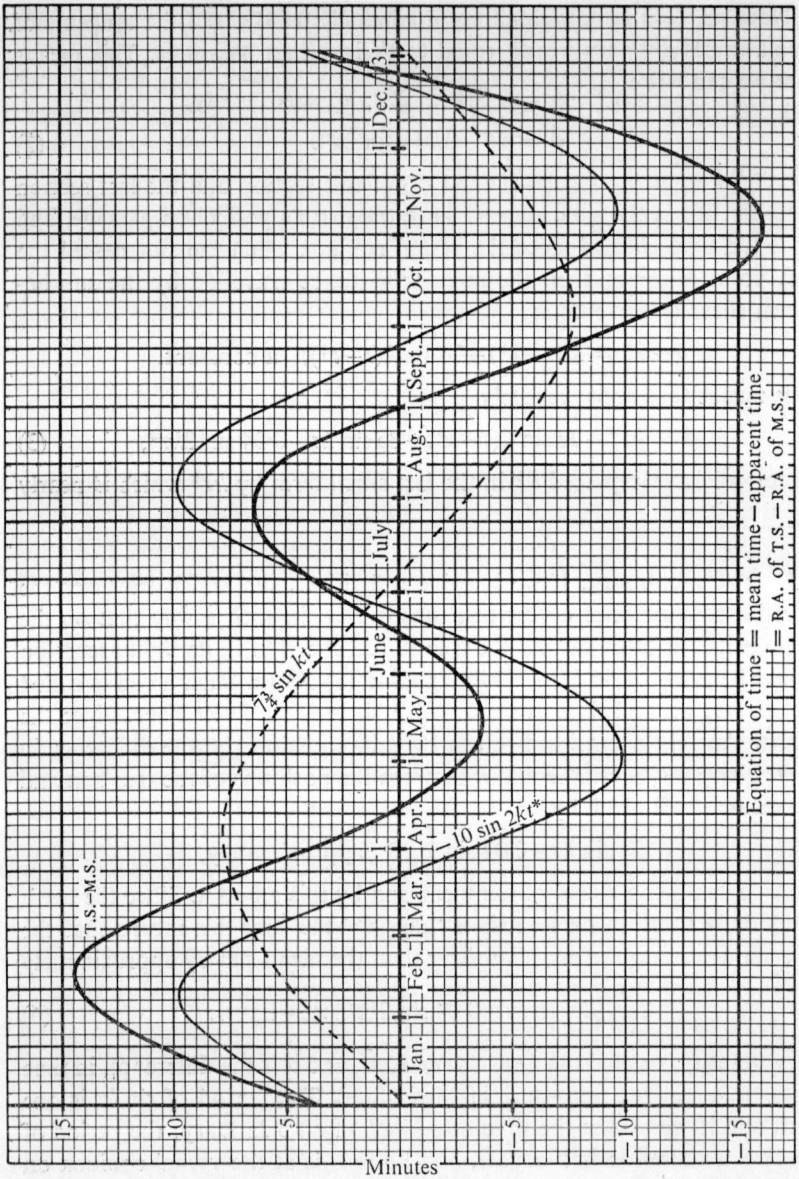

Fig. 17

Equation of time = mean time — apparent time
[ ] = R.A. of T.S. — R.A. of M.S.

$7\frac{3}{4}\sin kt$

$-10 \sin 2kt*$

T.S.—M.S.

1 Dec. 31

1 Nov.

1 Oct.

1 Sept.

1 Aug.

July

June

1 May

1 Apr.

1 Mar.

1 Feb.

1 Jan.

Minutes

15

10

5

5

10

15

For example, on 1 August the equation of time is about 6 minutes, so that the passage of the true sun across the Greenwich meridian is 6 minutes later than that of the mean sun. Furthermore, if the time is read off a sundial at Greenwich during that day, 6 minutes would have to be added to give Greenwich Mean Time.

The navigator, when checking his position by using the altitude of the sun above the horizon, makes an allowance for the equation of time. This is necessary in order to relate the true sun he has observed to the mean sun involved in timing the observation.

It will be noted that the use of M.S. for time-keeping is emphasized in phrases such as Greenwich Mean Time.

Exercise D

1. Calculate to the nearest minute the length of the sidereal day.

2. The sidereal month—i.e. the period of the moon's orbit round the earth relative to the distant stars—averages 27·32 days, but it is a very variable period. The moon revolves almost in the plane of the earth's orbit, and in the same sense. Calculate the length of the *synodic month*—the average time from one full moon to the next.

3. How much, in angular measure, does the moon gain, in its apparent eastward motion, on the sun each day?

4. On what day in the year is the earth's angular velocity about the sun greatest? Is its actual orbital velocity greatest on this day too?

5. On what days in the year is the equation of time changing most rapidly— i.e. when is the greatest day-to-day difference between the times when the sun crosses the meridian?

6. If Orion now sets at midnight, will it be visible after midnight next month?

7. Will Sirius rise earlier or later tomorrow than it does today?

8. At Exeter, 3° 34′ W, is the sun ever due south at midday? If so, when?

9. Answer Question 8 for Land's End, 5° 44′ W.

10. Patrick O'Houlihan of Ballynabog (8° 45′ W) has a sundial in his garden which the leprechauns have used as a dining-table. Advise him (a) how to set it, (b) how to calibrate it to agree with his watch on St Patrick's Day, assuming that his watch reluctantly keeps G.M.T. Is there any other day in the year when it will be correct?

3.6 Time systems.

Greenwich Mean Time (G.M.T.)

A clock which is regulated to span 24 hours in a mean solar day and adjusted to read 12.00 when the mean sun is 'crossing' the Greenwich meridian is said to be keeping G.M.T.

Standard Time (s.t.)

In all civilized parts of the world clocks are regulated to 24 hours to the day, but otherwise the time kept by a country (or part of a country) is a matter of national choice. It is called the standard time for that country and the correction in hours and minutes needed to convert it into G.M.T. is given in almanacs (such as *Whitaker's Almanac*).

British Standard Time (b.s.t.)

In this country we keep a standard time 1 hour ahead of G.M.T. and this is known as British Standard Time.

Local Mean Time (l.m.t.)

A clock keeps the Local Mean Time of a place if it is adjusted to read 12 o'clock (noon) when the mean sun crosses the meridian of that place.

Example 1. What is the difference in L.M.T. readings at places

$$A \{37° \text{ N, } 43° \text{ W}\} \quad \text{and} \quad B \{13° \text{ S, } 10° \text{ E}\}?$$

The difference in longitude is 53° and the sun's movement in longitude is at the rate of 24 hours for 360°, which is equivalent to 15° per hour or 1° in 4 minutes.

Therefore the time interval from the sun's meridional passage at B to its meridional passage at A is 3 hours 32 minutes. Hence the L.M.T. at B exceeds the L.M.T. at A by 3 hours 32 minutes. ('Exceeds' because when the sun is passing between the meridians of A and B, B's clock is after noon while A's clock is before noon.)

Example 2. What is the G.M.T. if the L.M.T. at a place in longitude 100° W is 19.30, 3 June (using the 24 hour clock system in which 19.30 is equivalent to 7.30 p.m.)?

G.M.T. is the L.M.T. on the Greenwich meridian.

Now $100° = 6 \times 15° + 10°$, so that G.M.T. will exceed the L.M.T. at 100° W by 6 hours 40 minutes. Therefore

$$\text{G.M.T.} = 19.30, \text{ 3 June} + 06.40 = 26.10, \text{ 3 June}$$

$$= 02.10, \text{ 4 June.}$$

Example 3. What is the difference between the Standard Time and the L.M.T. in Vancouver Is., British Columbia, on 18 January?

Vancouver Is. is (50° N, 126° W), so the L.M.T. there is 8 hours 24 minutes behind G.M.T. From *Whitaker's Almanac* their s.t. is 8 hours behind G.M.T. Hence the s.t. is 24 minutes ahead of L.M.T.

3.7   The date line. In general for places to the east of Greenwich near to the 180° E meridian, the time kept will be about 12 hours fast on G.M.T.; in places to the west of Greenwich near to the 180° W (which is the same meridian as 180° E), the time kept will be about 12 hours slow on G.M.T. Under such a time-keeping system, a ship sailing across the 180° meridian would need to adjust her ship's time by about a day—the date being increased if she is westbound, decreased if eastbound.

The adjustment need not be an exact 24 hours because a ship might, as a matter of convenience, have her clocks only 11 hours fast when to the west of the 180° meridian and perhaps 12 hours slow on the other side. Moreover, the adjustment need not be made at the instant of crossing this meridian but reserved maybe for the small hours of the night.

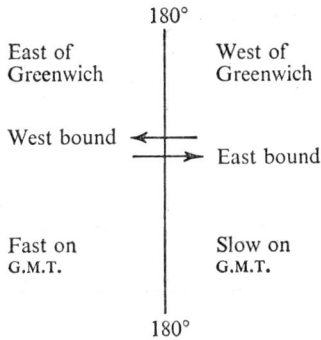

For places on land, lying near the 180° meridian, the standard times are approximately 12 hours fast or slow on G.M.T., but it must be remembered that standard times are chosen for convenience. The inhabitants of Eastern Siberia, for example, keep a time 13 hours fast on G.M.T. although parts of the group lie on one side and parts on the other side of the 180° meridian.

The 'date line' is a demarcation line running mainly down the 180° meridian, but with sweeps to the east and west as necessary to separate those places (mainly groups of islands) which are fast on G.M.T. from those which are slow on G.M.T. Places on either side of the date line will therefore differ in their time reckoning at any instant by about 24 hours.

Example 4. A ship leaves Tonga (20° S, 174° 30′ W) at 07.32, 5 May (Tonga S.T.) bound for Auckland (N.Z.) (36° 52′ S, 174° 48′ E). If the voyage takes 120 hours, at what time (New Zealand S.T.) will she arrive?

Tonga S.T. is 12.19 fast on G.M.T.
N.Z. S.T. is 12.00 fast on G.M.T.

Leaves 07.32	Tonga S.T.	5 May	
12.19			
19.13	G.M.T.	4 May	
Arrives 19.13	G.M.T.	9 May	
			(120 hours = 5 days)
12.00			
Arrives 07.13	N.Z. S.T.	10 May	

Example 5. An aircraft takes 3 hours to fly from Samoa (14° S, 172° W) to Fiji (18° S, 179° E). It leaves at 16.10, 3 June (Samoan S.T.); when does it arrive?

Leaves	16.10	3 June, Samoan S.T.
	11.00	Samoan S.T., 11 hours slow on G.M.T.
	03.10	4 June G.M.T.
Duration	05.00	
Arrives	08.10	4 June G.M.T.
	12.00	Fiji S.T., 12 hours fast on G.M.T.
Arrives	20.10	4 June, Fiji S.T.

Example 6. A radio signal is transmitted at 00.15 Tuesday (S.T.) from Wrangell Is. (71° N, 179° E). At what S.T. will it be received in the Savage Is. (18° 56′ S, 169° 52′ W)?

Transmitted	00.15	*Tuesday*, Wrangell Is. S.T.
	13.00	Wrangell Is. S.T., 13 hours fast on G.M.T.
	11.15	Monday G.M.T.
	11.20	Savage Is. S.T., 11.20 slow on G.M.T.
Received	23.55	*Sunday*, Savage Is. S.T.

Exercise E

1. How long does the sun take to move through a longitude of (i) 15°; (ii) 60°; (iii) 2°; (iv) 62°?

2. What is the difference in L.M.T. at A (43° N, 17° W) and B (20° S, 47° W)? If the L.M.T. at A is 09.30, what is it at B?

3. What is the difference in L.M.T. at P (10° N, 153° E) and Q (45° N, 70° E). If the L.M.T. at Q is 20.13, what is it at P?

4. If the L.M.T. in position (40° N, 116° E) is 12.00, what is the G.M.T.? Would it be reasonable for the S.T. at this place to be 8 hours fast on G.M.T.?

5. If the L.M.T. in position (65° S, 97° W) is 19.15 on 2 March, what is the G.M.T.?

6. What is the meridian of the place for which the L.M.T. is 05.12 at G.M.T. 18.35?

7. What is the difference, to the nearest minute, between B.S.T. and L.M.T. in Penzance (50° 06′ N, 5° 33′ W)?

8. An aircraft leaves Teheran at 10.00 Persian S.T. (3 hours 30 minutes fast on G.M.T.). What is the time of arrival in Bombay, in Indian S.T. (5 hours 30 minutes fast on G.M.T.), if the journey takes 4 hours?

412

9. If a flight from New York to Los Angeles takes 5 hours and the plane leaves at 16.40 Eastern s.t. (5 hours slow on g.m.t.), what is the time of arrival in Pacific s.t. (8 hours slow on g.m.t.)?

10. A ship leaves Singapore at 07.10 Singapore s.t. (7 hours 30 minutes fast on g.m.t.) on 6 June for Sydney. The journey takes 19 days 11 hours. Find the g.m.t. of departure and arrival, and hence the time of arrival in N.S.W. s.t. (10 hours fast on g.m.t.).

11. A wireless signal is sent from Christmas Is., Indian Ocean, at 15.20 7 March (s.t. 7 hours fast on g.m.t.). When will it be received in Christmas Is., Pacific Ocean (s.t. 9 hours slow on g.m.t.)?

12. If a flight from Honolulu to Wake Is. takes 5 hours 20 minutes, when, in Eastern Carolines s.t. (12 hours fast on g.m.t.), will it arrive if it leaves on 7 February at 17.30, Hawaiian s.t. (10 hours slow on g.m.t.)?

13. How many Sundays will be enjoyed during February by the crew of a ship that sails weekly from Tonga to Samoa, leaving Tonga on Sunday, 1 February 1976?

4. THE GREGORIAN CALENDAR

Our present calendar system was established by Pope Gregory XIII in 1582 and adopted in this country in 1752 (in Russia only in 1918). It superseded the regulation of the year introduced by Julius Caesar in 46 B.C.

The duration from the spring equinox, when the sun is at the first point of Aries, to the next spring equinox is called an equinoctial, or tropical year, and is approximately 365·2422 mean solar days. (Owing to precession this differs both from the sidereal year and from the interval between successive passages of the sun through perihelion.) The problem was how to regulate systematically the number of days in each year—and of course in any one year it must be a whole number of days—so that, over the span of centuries, the average number of days in a year would approximate closely to the length of a tropical year, thereby linking the calendar dates with the seasons.

The Gregorian rule is that a non-century year whose year number is divisible by 4 and a century year whose number is divisible by 400 are leap years; the others are common years. (Thus 1900 was not a leap year, whereas 2000 will be a leap year.) Hence there are 97 leap years and therefore 97 extra days in every 400 years. This will give as the average number of days in a year $365 + \frac{97}{400}$ which is 365·2425. The calendar year therefore exceeds a tropical year by only 0·0003 days (approximately), so that in 10 000 years the calendar will gain only some three days relative to the seasons.

It is proposed to improve the Gregorian rule by making the year 4000 and its multiples common years. The error would then not be more than one day in 20 000 years.

5. THE MOON

The earth's natural satellite, the moon, has a diameter of 3460 kilometres, and revolves eastwards around us once a month at a distance of about 380000 kilometres in an elliptically shaped orbit which is inclined to the ecliptic at approximately 5°. The same face of the moon is always showing towards the earth, so that the moon has a rotation period equal to its period of revolution.

The eccentricity of the orbit is about 0·055 and the maximum and minimum distances are 404340 kilometres and 338340 kilometres. The period of revolution relative to the stars, of about 27·32 days, is called a sidereal month, while the period relative to the sun is called a synodic month and lasts about 29·53 days; the latter is the interval between new moons. It can easily be checked by observation on successive nights that the moon moves about 13° to the eastward each day relative to the stars and therefore about 12° relative to the sun.

5.1 Phases of the moon. Figure 18 shows in the outer ring how the moon is illuminated by the sun during its revolution about the earth; the

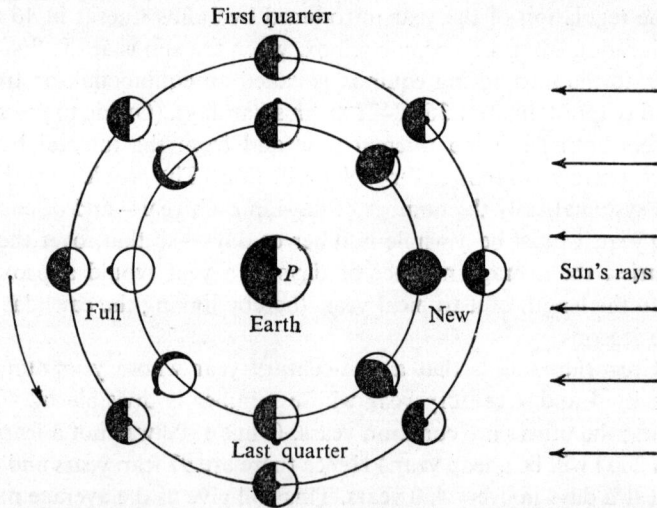

Fig. 18

inner ring shows how the moon, in its different phases, appears to us, and the dots mark points which correspond in the 'moons' on the two rings. It should also be noticed that for the 'moons' on the outer ring, only that part lying inside the ring will be visible from the earth. The positions of the moon at New, First Quarter, Full and Last Quarter are as shown.

414

In Figure 19 the sequence of changes shown previously in the inner ring of the Figure 18 is arranged in strip form to make the distinction between 'right' and 'left' more clear. It roughly depicts the appearance of the moon when it is to the south of us. It will be seen that the demarcation line between light and dark always moves across the moon from right to left as time goes on, and one can apply this rule to decide whether a moon of any particular shape is waxing or waning (the meaning of these terms as also of crescent and gibbous, is shown in Figure 19). If the moon crosses an observer's meridian to the north of him, then the demarcation line moves instead from left to right.

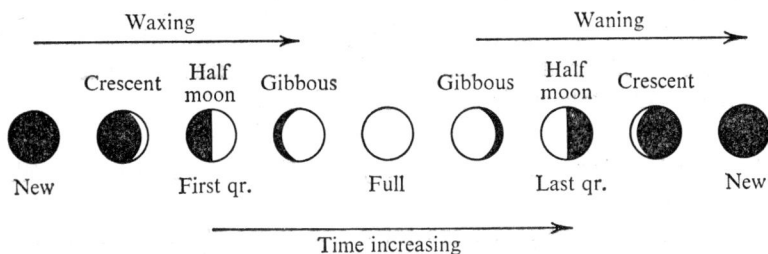

Fig. 19

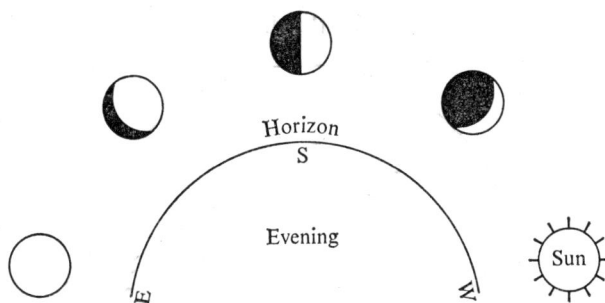

Fig. 20

The relative positions of the sun and moon in the sky for different phases is shown in Figure 20. A crescent-shaped moon must necessarily be 'close' to the sun in the sky, while a full moon will be in the opposite part of the sky. The orbits of the sun and moon both lie approximately in the same plane, namely the ecliptic. Hence, for example, in January a full moon will have roughly the same position on the celestial sphere as the sun has in July. Therefore a January full moon in the northern hemisphere climbs high in the observer's sky. A summer full moon in the northern hemisphere, on the other hand, keeps low down like a winter's sun.

415

6. PLANETS AND THEIR APPARENT MOVEMENTS

Some facts about the planets have been set out in the table at the beginning of this chapter. Apart from perturbations caused mainly by their attraction one on another, the true motions of the planets are of the simple form described in Section 1.1. We shall now make the approximations that the orbits are circular and lie in the plane of the ecliptic. However, the planets appear to us to have a very irregular movement across the sky—the word 'planet' is from the Greek for wanderer—but this is almost entirely due to the fact that we observe them from a moving base, the earth.

We shall deal separately with the inferior planets, whose orbits lie within the earth's orbit, and the superior planets which lie outside, as their apparent movements are in some ways different.

6.1   The inferior planets—Mercury and Venus—visible to the naked eye.

In Figure 21 E represents the earth and P the planet, each having a circular orbit around the sun, S. The star background we take as a sphere of extremely large radius centred on the sun. As E and P move round their orbits the position of P viewed from E and seen against the star background will be determined at any time by the direction EP.

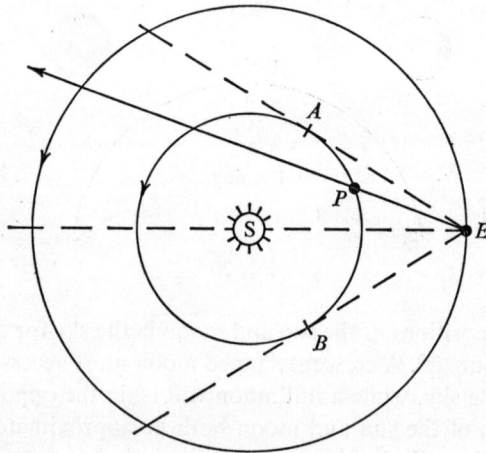

Fig. 21

Since P completes its orbit more quickly than E, the variation in direction EP is a backwards and forwards swing about the direction ES, which is itself changing steadily as E moves round.

416

The planet, therefore, has the general yearly movement of the sun around the ecliptic, on to which is imposed a swinging movement in the plane of the ecliptic to a maximum angular elongation equal to the angle SEA, to the right and left. The elongation for Venus is about 46° (the ratio $SA : SE$ is the ratio of the radii of the orbits of Venus and the earth—about 0·723, which is sin 46° 18′). In this country, when Venus is to the 'right' of the sun it will be seen as a 'morning star' and when to the left, as an 'evening star'. The arc of the orbit from A to B, behind the sun is longer than from B to A in front of the sun, so that the swing from right to left takes the greater time. To compute the time of a complete oscillation, we use the equation of synodic motion (Section 3.1). If E_s is the length of the sidereal year, V_s that of Venus's sidereal period, and V_e that of its *synodic* period—i.e. the period of a complete oscillation relative to the earth, then we have

$$\frac{1}{V_s} - \frac{1}{E_s} = \frac{1}{V_e}.$$

Inserting the values of E_s (365·26 days) and V_s (224·7 days), we obtain 584 days for V_e. The synodic period of Mercury can be calculated in the same way, and is left as an exercise.

6.2 The superior planets—Mars, Jupiter, Saturn (visible to the naked eye), Uranus, Neptune and Pluto.

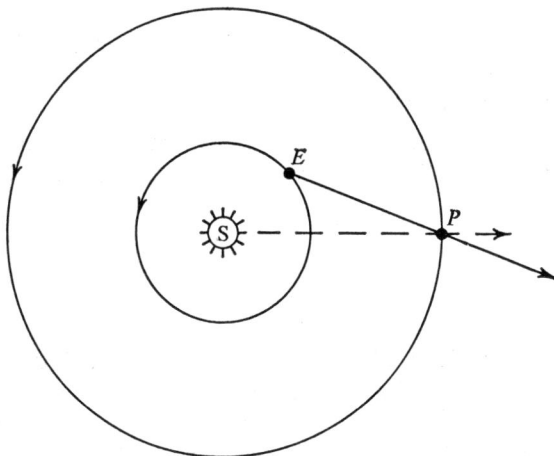

Fig. 22

Figure 22 is similar to Figure 21, but with the roles of E and P interchanged we see that the change in direction of EP is again a backwards and forwards swing, due now to the earth's movement, superimposed on the steady change in direction of SP due to the planet's motion in its orbit.

417

In this case, of course, the planet's elongation, $\angle SEP$, can have any value, and the superior planets, therefore, are not always near the sun in the sky.

6.3   Retrograde motion of the planets. The to and fro oscillation and the steady movement, which for the different planets are in different proportions, might have compounded into a motion which was at times retrograde (as for a climber whose general direction is upwards but who slips back periodically) or to one which is direct—varying speed but always in the same direction (as for a worm). In fact we shall now show that the motion is always partly retrograde.

Consider the special cases when the earth and the planet are in line with the sun, and let their orbital speeds be as shown in Figure 23.

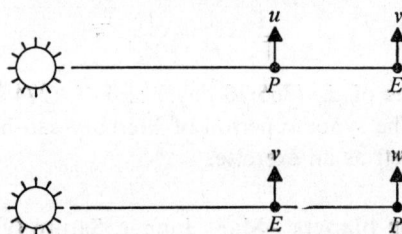

Fig. 23

If $u > v > w$, the segment EP will at this time be rotating clockwise, and this corresponds to a retrograde motion for the planet.

Now by Kepler's Third Law (see Section 1.4 and use the notation there) T^2/D^3 is constant for all planets. For the approximate circular orbits of circumference πD, the time T for one revolution at speed V is $\pi D/V$. Hence

$$T^2 = \frac{\pi^2 D^2}{V^2}.$$

Therefore
$$\frac{\pi^2 D^2}{V^2 D^3}$$

is constant, from which we have that $V^2 D$ is constant. This means that as the diameter of the orbit increases the speed of the planet round the orbit decreases—the further from the sun the slower the planet moves. Hence the conditions required for retrograde motion are satisfied in the particular cases considered.

6.4   Phases of the inferior planets. The figure, which is not to scale, shows the phases of an inferior planet. The inner circle shows the illuminated part of the planet facing the sun—for a planet is only visible by reflected light. The outer ring shows the appearance of the planet to us.

418

Since the apparent size of the planet will depend on its distance from us, as the planet 'waxes' its diameter will appear to decrease, and to increase again as it wanes. For this reason the brightness of either Venus or Mercury is not as variable as one might otherwise expect.

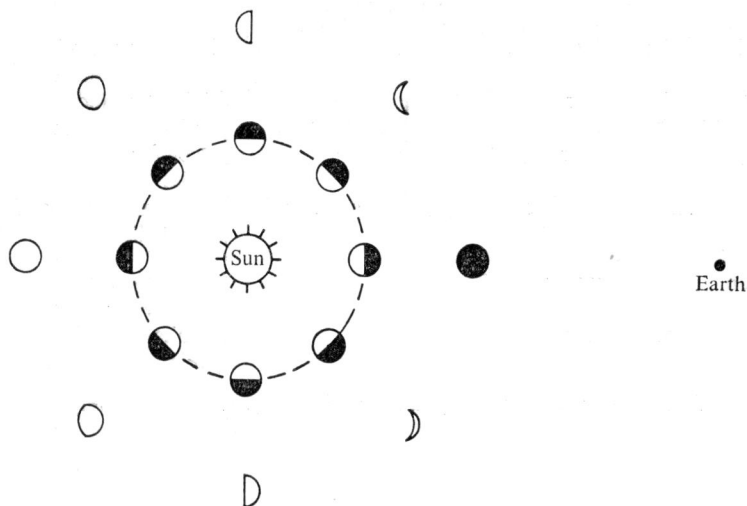

Fig. 24

6.5 Phases of a superior planet. Using the last figure with the earth now between the sun and the planet it will be found that only full or gibbous shapes are possible. The distant planets in fact show very little phase change.

Exercise F

1. What is the maximum elongation for Mercury (to the nearest degree)? Express this in terms of the spread between your outstretched thumb and little finger at full arm's length.

2. (i) Find the ratio of the orbital speeds (assuming circular orbits) of (a) Venus and the earth, (b) the earth and Mars.

(ii) Find the ratio of the distances at our closest approach to Venus and to Mars.

(iii) Compare the angular speed of retrograde motion for Venus when in conjunction with the sun, with that of Mars when in opposition to the sun.

3. Draw to scale the (circular) orbits of the earth and Mars. If the sun is now on the line joining the earth to Mars, show month by month for the next 18 months how the direction of Mars will change. Show this by marking the positions of Mars on the circumference of a circle representing the celestial ecliptic. (Assume that Mars lies in the ecliptic.)

4. Compare in general terms the apparent motions of Mars and Jupiter.

5. What is 'syzygy'?

Table of time-periods

(Mean solar days)

Earth	Sidereal day	23 h 56 min 4 s		Rotation period relative to stars
	Sidereal year	365·2564		Orbital period relative to stars
	Tropical year	365·2422		Orbital period relative to ♈
	Anomalistic year	365·2596	mean solar days	Between passages of perihelion
Moon	Sidereal month	27·3217		Orbital period relative to stars
	Synodic month	29·5306		Orbital period relative to earth

Period of precession of equinox = 25 800 years, approximately.

INDEX

acceleration, average, 194, 327–8, 354; as gradient of speed-time graph, 195; produced by a force, 335, 354; produced by gravity, 294, 338–9; and speed, 193–8; uniform (constant), 193, 216–17, 227–9, 328–34; variable, 217–20

action, 342; normal, 295, 300

addition, of vectors, 287–9

addition formulae, for sine and cosine, 306, 317–19; for tangent, 319

additivity of mass, 342–3

air speed, 291

alternating current, 235–6

altitude, of a celestial body, 400

angle measure, 237–41

angular velocity, 242–3; of earth, 243, 405; of planets about sun, 393

aphelion, 392

apogee, 392

approximations, linear, 213, 277; linear, to $\sin x$ and $\cos x$, 244–5, quadratic, 213

areas, under curves, 216–27, 281–3

Aries, first point of, 398–9, 402–3

axes, major, of planetary paths, 391; relation of period of revolution to, 391, 393

base vectors, 300, 323

bases of logarithms, 294–5

binomial coefficients, 379

binomial theorem, 378–9

calculus, 205

calendar, Gregorian, 413

celestial sphere, 397–403

chain rule, for derivatives of composite functions, 269

characteristic, in computation by logarithms, 252

Charlier's checks, on calculation of standard deviation, 362–3

circle, equation of, 310

circular functions, see sine and cosine

circular measure, 238–41

collision, momentum in, 350, 354

components, of a vector, 300–2, 305; of forces, 300, 344; scalar product in terms of, 306

composite functions, 267–8; derivative of, 268–70

computation, by iteration, 259–60; by means of logarithms, 248–53; of standard deviation, 359–65; of tables of logarithms, 254–5

computers, programming of, 255–9

contact forces, 295–6; normal, 295, 300, 342

cooling, Newton's law of, 280

coordinates, expressions in, 302–4

correlation coefficients, 365; rank, 365–8, 371

cosine rule, for solution of triangles, 312–14

cosine and sine, see sine and cosine

course and track, 290–3

cumulative frequency diagram, 383

curves, areas under, 216–27, 281–3

cycle, of alternating current, 236

date line, 411

day, mean solar, 404–5; sidereal, 420

decision boxes, in flow diagrams, 257

declination, 399

derivative, 199; of an area under a curve, 220–1; of composite functions, 268–70; of exponential functions, 275–7; of inverse functions, 274–5; of logarithmic function, 277; of powers of x, 202–4; of reciprocal function, 271–3

derived function, 199–200; differentiation as process of finding, 202

deviation, mean absolute, 357; from mean and median, 356–7; standard, 358–65

differentiation, 193–215; applications of, in mathematics, 208–13, and physics, 205–6; notation for, 204–5; see also derivative

displacement, 324; with uniform acceleration, 329–31

distance, as area under velocity-time graph, 216–20; astronomical unit of, 394; expressed in coordinates, 302

division, by use of logarithms, 251–3

dot product, of vectors, 304–7

drift, 291

dynamics, 322–54

earth, attraction of, 294, 338; orbit of, 395–6; seasons on, 396; variable angular velocity of, 405

eccentricity of an ellipse, 391

ecliptic, celestial, 396, 398, 401; inclination of, 406–7; plane of, 390

ellipse, construction of, 391; equation of, 392

variance, 360

variation, coefficient of, 364

vectors, 287–321; addition formulae for, 306, 317–19; components of, 300–2, 305; definition of, 287; multiplication of, by numbers, 297–9; perpendicular, 307–8; scalar product of, 304–7, 331–4

velocity, 324–7; angular, 242–3; areal, of planets, 391, 392–3; average, 324, 354; components of, 301, 325–6; constant, 216; and momentum, 348; variable angular, of earth, 405; *see also* speed

velocity-time graph, distance travelled as area under, 216–20, 228

weight, 294, 338, 343

wind velocity, 291

year, anomalistic, 420; equinoctial or tropical, 413, 420; sidereal, 404, 420

zero vector, 288

zeros, working, 357, 360

zodiac, 396–7, 403